AF566885

Reviews and critical articles covering the entire field of normal anatomy (cytology, histology, cyto- and histochemistry, electron microscopy, macroscopy, experimental morphology and embryology and comparative anatomy) are published in Advances in Anatomy, Embryology and Cell Biology. Papers dealing with anthropology and clinical morphology that aim to encourage cooperation between anatomy and related disciplines will also be accepted. Papers are normally commissioned. Original papers and communications may be submitted and will be considered for publication provided they meet the requirements of a review article and thus fit into the scope of "Advances". English language is preferred, but in exceptional cases French or German papers will be accepted.

It is a fundamental condition that submitted manuscripts have not been and will not simultaneously be submitted or published elsewhere. With the acceptance of a manuscript for publication, the publisher acquires full and exclusive copyright for all languages and countries.

Twenty-five copies of each paper are supplied free of charge.

Advances in Anatomy
Embryology and Cell Biology

Vol. 174

Springer-Verlag Berlin Heidelberg GmbH

S. Geyer

The Microstructural Border Between the Motor and the Cognitive Domain in the Human Cerebral Cortex

With 25 Figures and 3 Tables

Dr. Stefan Geyer

C. & O. Vogt Brain Research Institute
University of Düsseldorf
Universitätsstr. 1
40225 Düsseldorf
Germany

e-mail: stefan@hirn.uni-duesseldorf.de

ISSN 0301-5556

Library of Congress Cataloging-in-Publication Data

Geyer, Stefan, 1963–
The microstructural border between the motor and the cognitive domain in the human cerebral cortex / Stefan Geyer.
p. cm. – (Advances in anatomy, embryoloy, and cell biology; v. 174)
Includes bibliographical references and index.

DOI 10.1007/978-3-642-18910-4
1. Motor cortex–Cytology. 2. Cerebral cortex–Cytology. 3. Cytoarchitectonics. I. Title. II. Series.

springeronline.com

Originally published by Springer-Verlag Berlin Heidelberg New York in 2004
MyCopy version of the original edition 2004

Typesetting: Stürtz, Würzburg

Printed on acid-free paper 27/3150/Ag – 5 4 3 2 1 0

springer.com/mycopy

"Heureka!"
Archimedes of Syracuse
(c. 285–212 BC)

Contents

1 Prologue: Toward the Concept of a Cortical Control of Voluntary Movements

1.1 Jackson and the Concept of a Motor Cortex

In 1863 John Hughlings Jackson wrote about unilateral epileptic seizures caused by syphilis: "In very many cases of epilepsy, and especially in syphilitic epilepsy, the convulsions are limited to one side of the body; and, as autopsies of patients who have died after syphilitic epilepsy appear to show, the cause is obvious organic disease on the side of the brain, opposite to the side of the body convulsed, frequently on the surface of the hemisphere" (cited from Finger 1994; Jackson 1863). In the middle of the nineteenth century the concept of a cortical control of voluntary movements that is fully accepted today was hotly debated. It was widely believed that the corpus striatum was the uppermost center of motor control and the origin of the descending motor tracts. Jackson, on the other hand, was intrigued by the spread of seizures over specific body parts, e.g., from the hand to the arm to the face on one side of the body or from the hip to the leg to the foot and toes. Most frequently affected were the muscles of the face, hand, and foot. Only in severe cases did the seizure cross the midline and involve the entire body, usually followed by loss of consciousness. In hemiplegic patients Jackson observed that unilateral seizures were most likely to affect those parts of the body that were also most affected in hemiplegia. It appeared to him as if epilepsy were the "mobile counterpart of hemiplegia". Based on these observations, Jackson proposed a sector of the cerebral cortex with motor functions and topographically organized representations of muscles. In this motor cortex, those body parts most frequently affected by epileptic seizures (i.e., face, hand, and foot) were assumed to have larger spatial representations than body parts less frequently affected.

1.2 Experimental Confirmation of the Motor Cortex in Laboratory Animals

Jackson's concept was based on clinical observations and brilliant conclusions. However, he could not confirm experimentally the existence of such a motor cortex, nor could he define its precise topography. This succeeded a few years later. Several times during the nineteenth century, scientists had attempted to induce body movements by electrically stimulating the cortex of laboratory animals. The experiments, however, were not successful or provided conflicting results. In 1870, Gustav Fritsch and Eduard Hitzig stimulated the cortex of dogs—not in a laboratory, but on a dressing table in a bedroom of Hitzig's house in Berlin. They found discrete regions of the cortex that gave muscular responses in the forepaw, hindpaw, face, and neck on the opposite side. A unilateral lesion of the forepaw area with a scalpel handle significantly impaired motor performance but did not cause sensory deficits (Fritsch and Hitzig 1870). Fritsch and Hitzig interpreted these findings as evidence for a discrete motor cortex with a somatotopic (i.e., topographically correct) representation of the body.

Several years later David Ferrier replicated the findings of Fritsch and Hitzig in primates. By stimulating the cortex rostral to the central sulcus with faradic current he obtained a broad spectrum of movements. A monkey with a small motor cortex lesion developed paralysis restricted to the opposite hand and forearm but no deficits in touch and pain sensation. Another with a lesion that encompassed large parts of the left frontal, parietal, and temporal lobe displayed a right hemiplegia. When the brain was examined histologically, Ferrier found a degeneration of the pyramidal tract extending as far as the lumbar spinal cord (Ferrier 1876, 1886; Ferrier and Yeo 1885). In 1881, he even took a monkey with a unilateral lesion of the left motor cortex to the Seventh International Medical Congress in London. When the hemiplegic animal, operated upon seven months earlier, limped into the demonstration room, Jean-Martin Charcot remarked: "It is a patient!" (cited from Finger 1994). Based on the observation that patients with cortical lesions showed deficits predominantly in voluntary movements whereas involuntary and "automatic" movements were less affected, Jackson had already hypothesized that the cerebral cortex was responsible for voluntary actions whereas subcortical centers played a role in more "automatic" activities. Ferrier confirmed these conclusions with observations in laboratory animals. Primates with a more elaborate cortex rely, when interacting with their environment, more on voluntary movements than do animals lower on the phylogenetic scale, e.g., reptiles or amphibians. In other words, the higher an animal's level on the phylogenetic scale, the more its voluntary movements are affected by damage to the motor cortex. The findings further supported the concept that voluntary movements, being phylogenetically more "modern," are represented mainly on a cortical level.

Around the turn of the twentieth century, Sidney Grünbaum and Charles Sherrington supplemented Ferrier's studies with stimulation and lesion experiments in apes. In accordance with Ferrier's findings, they reported that stimuli that evoked motor responses from the precentral gyrus were ineffective when applied in the postcentral gyrus. Likewise, ablations of the precentral gyrus resulted in paralysis, whereas lesions in the postcentral gyrus did not (Grünbaum and Sherrington 1902, 1903). Grünbaum and Sherrington's carefully conducted experiments and meticulously published data remained seminal throughout the twentieth century.

1.3 Maps of the Motor Cortex in Humans

The first comprehensive map of the human motor cortex is usually attributed to the Canadian neurosurgeon Wilder Penfield. The sketch of its somatotopic organization (Penfield and Rasmussen 1952)—better known as the "motor homunculus"—can be found in most neuroscience textbooks. However, at the beginning of the twentieth century, Fedor Krause in Berlin had already attempted to map the human motor cortex. Working with patients who underwent surgery for epilepsy, he stimulated the exposed precentral gyrus with faradic current and made records of the induced movements. His map of the somatotopic organization of the motor cortex appeared in 1911 in his book *Chirurgie des Gehirns und Rückenmarks nach eigenen Erfahrungen*, anticipating by some decades most features of Penfield's "homunculus." Unfortunately, Krause's contributions are largely forgotten today.

1.4 Differences in Function Between Motor and Premotor Cortex

At the turn of the twentieth century, new impulses emerged from the analysis of the microanatomy of the cortex, i.e., the neurons (cytoarchitecture) and the myelin sheaths surrounding the neuronal processes (myeloarchitecture). Several of these microanatomical brain maps showed that the cortical area on the precentral gyrus ["precentral area" according to Campbell (1905) or "area 4" according to Brodmann (1909)] differed in cytoarchitecture from the rostrally adjoining area ("intermediate precentral area" according to Campbell or "area 6" according to Brodmann). In accord with the elementary biological concept: "What differs in structure should also differ in function (and vice versa)," the question that came up immediately was: "Is there another area rostral to the motor cortex with different motor properties?"

Some years earlier David Ferrier (1886) and Eduard Hitzig (1904) had noted that electrical stimulation of a region in front of the precentral gyrus in primates and dogs resulted in movements of the head and eyes to the opposite side. In 1931, Otfrid Foerster also reported that in humans stimulation of a comparable area caused the eyes to move to the contralateral side.

John Fulton, Paul Bucy, and Cécile and Oskar Vogt conducted experiments to further characterize the functional properties of the "intermediate precentral area" or "area 6" in nonhuman primates. The work of Foerster led to a better understanding of the differences between motor and premotor cortex in humans. Electrophysiological studies showed that stimulation of area 6 induced discrete movements on the opposite side of the body. However, in area 6 more current was needed to elicit responses compared to area 4. Since efferent projections from area 6 may terminate in area 4 and from there descend to the spinal cord, damage to area 4 severely impaired movements upon stimulation of area 6. Conversely, a lesion in area 6 did not affect movements upon stimulation of area 4 (Bucy 1934; Foerster 1931; Vogt and Vogt 1919). A lesion involving both area 4 and area 6 led to spasticity. Symptoms differed, however, when only one area was damaged. A lesion limited to area 6 was associated with spasticity. The degree of spasticity increased, whereas the degree of recovery decreased, as one ascended the phylogenetic scale. A lesion limited to area 4 led to a flaccid paralysis (Fulton 1937).

Early in the 1950s, Wilder Penfield and Keasley Welch described in humans another ("supplementary") motor area (SMA) on the mesial (i.e., facing the interhemispheric fissure) aspect of area 6 rostral to the leg representation of the primary motor cortex (area 4). By electrically stimulating the cortex during neurosurgical operations, Penfield and Welch obtained motor responses on the contralateral side: arm and leg movements more complex than those induced by stimulation of the primary motor cortex, movements of the head and eyes to the opposite side, and movements of the face and tongue as if the patient was attempting to speak. In contrast to the primary motor cortex, a somatotopic map of the body was not found. Lesions of the SMA did not lead to any persisting motor deficits (Penfield and Welch 1951). One year later, Clinton Woolsey and coworkers described a comparable area in nonhuman primates at a corresponding position in the frontal lobe. When the cortex was electrically stimulated, movements similar to those in humans were obtained that were also more complex than those evoked from the primary motor cortex, i.e., involving several joints or slow and tonic movements. In contrast to humans, however, Woolsey found a somatotopic organization in the nonhuman SMA (Woolsey et al. 1952).

Until the second half of the twentieth century, these "classical" findings supported the concept of a subdivision of the cortical motor system into three regions: the primary motor cortex (area 4) in the caudal wall and on the vertex of the precentral gyrus, the supplementary motor area further rostral on the mesial aspect of the frontal lobe (mesial part of area 6), and the premotor cortex on the lateral aspect of the frontal lobe (lateral part of area 6).

New findings in nonhuman primates, however, have shown over the last years that this relatively simple view is no longer adequate.

1.5 Cytoarchitectonic Analysis of the Motor and Premotor Cortex

Around the turn of the century, a group of anatomists, e.g., Alfred Walter Campbell, Grafton Elliot Smith, Korbinian Brodmann, and Cécile and Oskar Vogt set out to study the microanatomy of the cerebral cortex. They noticed that the size, shape, packing density, and lamination of neurons (cytoarchitecture) and the distribution pattern of myelin sheaths (myeloarchitecture) were not uniform across the cerebral cortex. Instead, there were marked regional variations. This allowed the delineation of cortical regions or areas, characterized by a uniform cyto- or myeloarchitectonic pattern, and the definition of borders between areas where the architectonic pattern changed. In the human cortex, Brodmann (1909) defined approximately 50 areas, and the Vogts (1919) delineated more than 200 regions. In keeping with the biological concept that differences in structure should reflect differences in function (and vice versa), the question arose: "What is the functional meaning of these areas?" In his famous book *Vergleichende Lokalisationslehre der Großhirnrinde*, published in 1909, Brodmann wrote: "The polymorphism of cells, their asynchronous histogenetic differentiation, the strict regional separation of certain cell types, and the regular appearance of similar (homologous) cell types in identical positions on the cortical surface in all mammals, all justify the view that in the cerebral cortex a broad division of function has occurred among the cellular elements, in other words that functional specificity is divided between cells of different individual morphology, different localisation and different capacity" (English translation in Garey 1994). Among the first investigators to provide experimental evidence in support of this hypothesis were the Vogts who, in 1919, published an article entitled "Allgemeinere Ergebnisse unserer Hirnforschung." They compared electrophysiological stimulation sites with the cytoarchitectonic pattern in nonhuman primates and found that the response properties of neurons changed across an architectonic border. In other words, architectonic areas were also functional entities. This triggered a "golden age" of cyto- and myeloarchitectonic mapping studies at the beginning of the twentieth century. Within two decades, parcellations of the cortex of humans and other mammals were published, to wit, the maps of Campbell (1905), Smith (1907), Brodmann (1909), the Vogts (1919), and von Economo and Koskinas (1925). In most maps, the caudal sector of the frontal lobe was subdivided into two areas: "precentral area" and "intermediate precentral area" according to Campbell (1905), areas 4 and 6 according to Brodmann (1909), and areas FA and FB according to von Economo and Koskinas (1925). Only the Vogts (1919) deviated from this concept and defined in their map of *Cercopithecus* three entities: areas 4, 6aα, and 6aβ.

1.6 Lashley and Clark's Criticism of Cytoarchitectonic Maps

This "localizationist" position, supported by the results of cyto- and myeloarchitectonic mapping studies, was challenged by a "holistic" way of viewing brain functions. The proponents of this concept, influenced by the ideas of the "Gestalt" psychology, argued that higher-level cognitive abilities (e.g., learning, memory, or intelligence) are functions of the entire brain and not of single cortical areas or modules. In support of this concept, Karl Lashley and Shepherd Ivory Franz conducted experiments in rats and showed that animals previously trained on a motor task showed no appreciable loss of this ability when various regions of the cortex (including large parts of the frontal lobe) were damaged (Franz and Lashley 1917; Lashley and Franz 1917). The post-lesional deficit seemed to be a function predominantly of lesion size and not of lesion location. It was not surprising that the proponents of "holism," especially Lashley, also criticized the validity of cytoarchitectonic mapping studies. In a study published in 1946, Lashley examined the cortex of a New World monkey (*Ateles geoffroyi*) and made a cytoarchitectonic map. George Clark, a colleague of Lashley, examined another *Ateles* brain and did the same. When the two investigators compared their parcellations, they found some agreement, but on the other hand the differences were such "as to cast doubt upon the reliability of both and to rule out either as a satisfactory guide to experimental work" (Lashley and Clark 1946). Upon closer examination of the two maps, they found that the differences were due to biological variability and also to each investigator's subjective criteria. The authors concluded that "in planning experimental work on the cortex it is desirable to have some guide to probable functional units in order that the work should not be undertaken wholly at random. For this purpose, however, the 'ideal' architectonic chart is nearly worthless, because individual variation is too great to make the chart significant for a single specimen, because the areal subdivisions are in large part anatomically meaningless, and because they are misleading as to the presumptive functional divisions of the cortex. ... The charting of areas in terms of poorly defined and variable characters, in the hope that future physiologic studies may some time reveal their significance, has contributed nothing to knowledge of cerebral organization and gives no promise of better achievement in the future" (Lashley and Clark 1946).

As a consequence, Percival Bailey and Gerhardt von Bonin in their cytoarchitectonic maps of the rhesus monkey (*Macaca mulatta*; von Bonin and Bailey 1947) and man (Bailey and von Bonin 1951) rejected the concept of subdividing the cortex into smaller and smaller areas and "haarscharfe Grenzen" (i.e., boundaries as sharp or fine as a hair; to paraphrase Oskar Vogt) between them. Instead, they interpreted—especially in their map of *Homo*—the six-layered isocortex as a uniform and ubiquitous cytoarchitectonic pattern (homotypical isocortex) with only a few regional variations (koniocortex, parakoniocortex, agranular cortex, agranular gigantopyramidal cortex, dysgranular cortex).

From a historical point of view, Lashley and Clark's criticism may seem plausible. From today's perspective, however, it was a tenet too radical and dogmatic. Yet, the two investigators formulated two points that the opponents of cytoarchitectonics have repeatedly raised over the years when arguing that cytoarchitectonic mapping is a technique too vague and unreliable to be termed "scientific."

First, when comparing their two *Ateles* maps, Lashley and Clark noticed that while differences between them were due to biological variability, each investigator's subjective criteria caused greater difference. The authors wrote: "A few of the criteria used in parcellation of the cortex are objective and precise, but the majority are stated in terms so vague as to be descriptively meaningless. Such terms as "rather large," "numerous," "loosely constructed," "feebly developed," are relative to standards which vary from one investigator to another and are not defined in more precise terms" (Lashley and Clark 1946). The variability in terms of number, size, and topography of areas among the maps published by different authors is obvious. To what extent this variability is a consequence of biology or subjectivity is open to discussion since there is no unique solution to a mathematical equation with two unknown variables.

The second point focuses on the uselessness of the "ideal" architectonic map, "because individual variation is too great to make the chart significant for a single specimen" (Lashley and Clark 1946). All "classical" maps (cf. above) were published in print format. Only with great uncertainty can structural data from these maps be matched with functional data from different brains [e.g., from electrophysiological studies or noninvasive imaging techniques like positron emission tomography (PET) or functional magnetic resonance imaging (fMRI)]. Furthermore, "classical" maps are schematic drawings that reflect the topographical situation in *one* representative brain and do not address the problem of interindividual macro- (see, e.g., Ono et al. 1990) and microanatomical (see, e.g., Rademacher et al. 1993) variability. In addition, these maps are "rigid" and not based on a spatial reference system which means they cannot be adapted to an individual brain. Multimodal integration of structural and functional data is impossible. More recently published atlases, e.g., the reference system of Talairach and Tournoux (1988), are of limited value as well, as their cortical maps are not based on genuine microstructural data. Instead, Talairach and Tournoux (1988) seem to have transferred each area from Brodmann's schematic drawing to a corresponding position on the cortex of their reference brain. In addition, the authors give only the approximate position of an area (borders between areas are not indicated) nor do they address the problem of interindividual variability (only one brain is depicted in the atlas). Noninvasive imaging techniques like PET or fMRI can relate foci of activation only to macroanatomical landmarks (i.e., gyri and sulci). Plenty of evidence in macaques and other nonhuman primates, however, has shown that it is microstructure—and not macroanatomy—that parallels function. Unfortunately, most microstructurally defined interareal borders in the human cortex do not match macroanatomical landmarks and they are topographically quite variable across different individuals (Amunts et al. 1999, 2000; Geyer et al. 1996, 1999; Rademacher et al. 1993; Rajkowska and Goldman-Rakic 1995; Roland et al. 1997; Roland and Zilles 1994, 1996; White et al. 1997). Hence, structural-functional correlations based only on macroanatomy are questionable and may account for at least some of the conflicting results functional imaging studies have provided in recent years, e.g., the debate whether (Hallett et al. 1994; Leonardo et al. 1995; Porro et al. 1996; Roth et al. 1996; Sabbah et al. 1995; Stephan et al. 1995) or not (Decety et al. 1994; Parsons et al. 1995; Rao et al. 1993; Roland et al. 1980; Sanes 1994) the human primary sensorimotor cortex is activated during imagined movements.

1.7
Observer-Independent Cytoarchitectonic Analysis and Probabilistic Microstructural–Functional Correlation— Two Responses to Lashley and Clark's Criticism

A response to Lashley and Clark's first criticism is quantifying cytoarchitectonic parameters and detecting borders between areas in an observer-independent fashion. Since a cortical area is characterized by a specific laminar pattern of cytoarchitectonic features, this pattern may be quantified by profile curves of cell density. To this end, histological sections are stained for cell bodies, digitized, and equidistant density profiles are extracted from the data matrix perpendicularly to the cortical layers from the pial surface to the border between gray and white matter. In order to compensate for variations in cortical thickness, the profiles are standardized in length and ten numerical parameters describing the profile's shape are then extracted from each curve. Comparing these parameters from two adjacent groups of profiles with multivariate statistical techniques detects significant differences in profile shape between the two groups. These differences may be interpreted as significant changes in the laminar cytoarchitectonic pattern or as borders between areas detected in an observer-independent way. Thus, the two unknown variables of the mathematical equation as suggested above may be reduced to *one*, namely, biological variability.

A response to Lashley and Clark's second criticism is to bring cytoarchitectonic data into a format that takes into account individual variability and may be adapted to another brain in order to directly compare structural and functional data. A recent development, namely, computerized brain atlases (Roland and Zilles 1994), offer the computational tools that are necessary to achieve this goal. On the one hand, genuine microstructural data (from observer-independent cytoarchitectonic analysis of cell-stained whole brain sections obtained from postmortem brains) are brought into the standard anatomical format of a computerized atlas. The degree of individual variability may be assessed by importing microstructural data from several brains. On the other hand, functional imaging data are brought into the identical standard anatomical format. Both data sets can then be superimposed and compared with each other on a probabilistic basis. This approach, termed *probabilistic microstructural-functional correlation* allows (1) definition of volumes of interest (VOIs) of cortical areas that are not based on macroanatomical landmarks but instead on cytoarchitectonic mapping of postmortem brains, and (2) determination in these VOIs of changes in regional cerebral blood flow during motor, sensory or cognitive tasks.

1.8 Goal of this Study

The caudal border of the motor cortex between primary motor area 4 and primary somatosensory area 3a lies close to the fundus of the central sulcus. The position of this border coincides with a macroanatomical landmark—an exception to the rule that micro- and macroanatomy do not correlate. No such coincidence has been found for the borders between area 4 (primary motor cortex) and area 6 (supplementary motor area and premotor cortex), or between area 6 and the prefrontal cortex. Unfortunately, the "classical" cytoarchitectonic maps are not precise enough to be used as valid guides to the topography of these borders. A probabilistic map of area 4 with its borders with area 3a and area 6 in standard anatomical format does exist (Geyer et al. 1996). What is missing is a map of area 6. The functional significance of the rostral border of area 6 with the prefrontal cortex, however, is pre-eminent: this border separates the "motor domain" of the supplementary motor and premotor cortex from the "cognitive domain" of the prefrontal cortex. Hence, it is the goal of this study to define in an observer-independent way the rostral border of area 6 in ten postmortem brains and to generate a probabilistic map of area 6 in the standard anatomical format of a computerized brain atlas.

2 Materials and Methods

The histological processing of postmortem brains, observer-independent cytoarchitectonic analysis, and the 3-D reconstruction and spatial normalization of the histological volumes are summarized as a flowchart in Table 1.

Table 1 Histological processing of postmortem brains, observer-independent cytoarchitectonic analysis, 3-D reconstruction, and spatial normalization of the histological volumes

Ten postmortem human brains, fixed in formalin or Bodian's fixative for several months

↓

T1-weighted MR scan (1.5 T Siemens scanner; 3-D FLASH sequence) (→ MR volume)

↓

Dehydrating in graded alcohols and embedding in paraffin

↓

Serial whole brain sections (20 μm thick; coronal or sagittal plane)

↓	↓
Histological sections	Images of paraffin blockface
↓	&
Mounting on coated slides	Digitized histological sections
↓	&
Staining with modified silver method	MR volume of the same brain
↓	↓
Subjective and objective cytoarchitectonic analysis	
↓	
Delineation of area 6	
— Transfer →	3-D reconstruction of the histological volume

↓

Spatial normalization of the 3-D reconstructed histological volume and area 6 to the reference brain of the computerized atlas

↓

Superimposition of normalized histological volumes and areas in 3-D space

↓

Population map of area 6

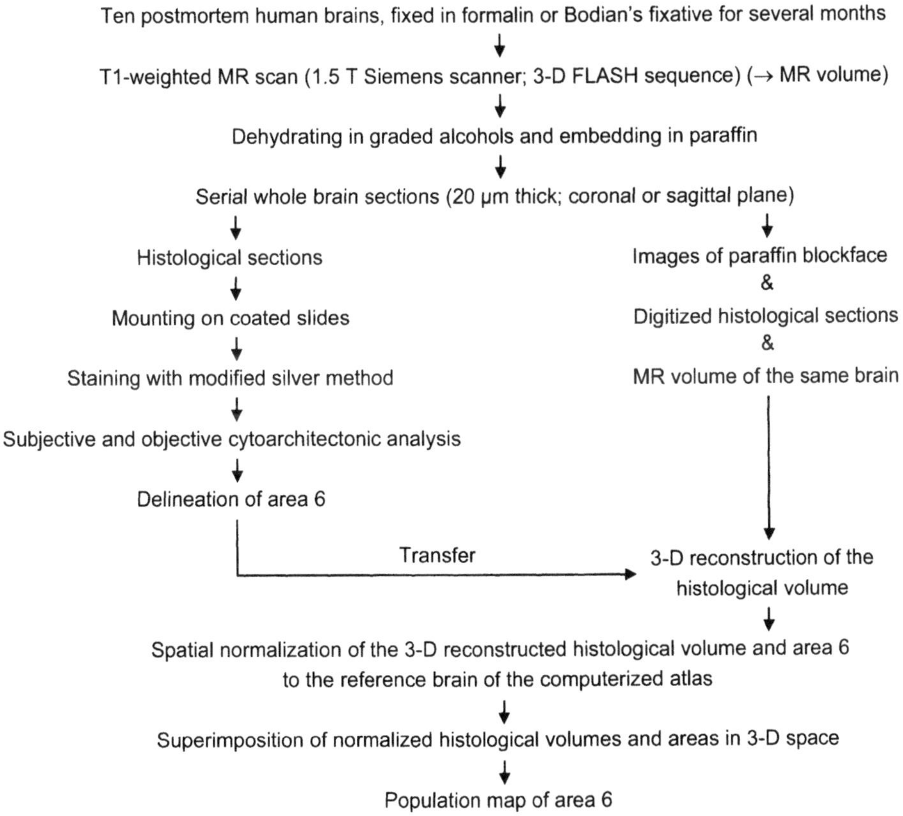

Table 2 Postmortem brains used for cytoarchitectonic analysis

Brain no.	Sex	Age (years)	Cause of death	Post-mortem delay (hours)	Fixative	Plane of sectioning
340/83	Female	79	Heart failure	<60	Formalin	Sagittal
646/79	Male	76	Right heart failure	24	Formalin	Sagittal
207/84	Male	75	Toxic glomerulonephritis	24	Formalin	Coronal
544/91	Female	79	Carcinoma of the bladder	24	Bodian	Coronal
189/92	Male	55	Carcinoma of the rectum	24	Formalin	Coronal
146/86	Male	37	Right heart failure	24	Formalin	Coronal
56/94	Female	72	Renal failure	12	Formalin	Coronal
2/95	Female	85	Mesenteric artery infarction	14	Bodian	Coronal
68/95	Female	79	Cardiorespiratory insufficiency	16	Bodian	Coronal
16/96	Male	54	Myocardial infarct	8	Formalin	Coronal

2.1 Histological Processing of Postmortem Brains

The cytoarchitectonic analysis was done in ten postmortem brains obtained at autopsy from subjects with no known history of neurological or psychiatric diseases (5 males and 5 females, mean age 69.1 years, range 37–85 years, cf. Table 2). All brains were obtained through the body donor program of the Department of Anatomy, University of Düsseldorf, Germany. The brains were suspended at the basilar artery and fixed in 4% formaldehyde or in Bodian's fixative (90 ml of 80% ethanol, 5 ml of 37% formaldehyde, and 5 ml of glacial acetic acid) for 5 months. After fixation, the arachnoidea was removed and T1-weighted magnetic resonance (MR) scans [MR volume; 1.5 T Siemens Magnetron SP scanner; 3-D fast, low angle shot (FLASH) pulse sequence; flip angle 40°; TR 40 ms; TE 5 ms; voxel size 1.17 mm (mediolateral) x 1 mm (rostrocaudal) x 1 mm (dorsoventral)] were acquired for documentation of brain size and shape before subsequent histological processing. The brains were dehydrated in graded alcohols, embedded in paraffin, and sectioned coronally or sagittally (20 µm whole brain sections) with a microtome for large sections. Images of the paraffin blockface were obtained after each 60th section (coronal plane) or 45th section (sagittal plane) with a CCD camera (XC-75; Sony; image matrix 256×256 pixels; 8 bit gray value resolution). Each 60th section (coronal plane) or 45th section (sagittal plane) was mounted on a gelatin-coated slide and stained for cell bodies with a modified silver method (Merker 1983) that yields high contrast between cell bodies and neuropil, a prerequisite for observer-independent cytoarchitectonic analysis.

2.2 Observer-Independent Cytoarchitectonic Analysis

In each cell-stained histological section, a rectangular region of interest (ROI) was defined and marked on the coverslip. These ROIs were scanned in a mosaic-like sequence with a digital camera (XC-75; Sony) attached to a microscope (Universal microscope; Zeiss) equipped with a computer-controlled motorized stage. The images (512×512 pixels covering a field of view 540×540 μm; 8 bit gray value resolution) were further processed with an image analysis software package (KS 400; Zeiss). In each image, the areal fraction of darkly stained cell bodies was calculated after adaptive thresholding (Schleicher and Zilles 1990) in square, adjoining fields, each 20 μm wide. This areal fraction [gray level index (GLI)] is highly correlated with the volume density of neurons (Wree et al. 1982). Glial and endothelial cells are equally distributed throughout the layers of the cortex, and are thus an additive—but constant—contribution to the volume density of neurons. The GLI was determined in each field (20×20 μm) within each image (540×540 μm); the resulting data matrix covering the entire ROI is the GLI image. Variations in staining intensities of cell bodies and/or background do not influence the GLI image because the GLI is extracted from binary images [with black cell bodies (gray value=0) and white neuropil (gray value=255)]. For further details see Schleicher and Zilles 1990.

Equidistant density profiles (125 μm wide; 200 μm spacing between adjacent profiles) covering the entire ROI were extracted from the GLI images perpendicularly to the cortical layers. The profiles extended from the border between layers I and II to the border between layer VI and the white matter. An outer (between layers I and II) and inner (between layer VI and the white matter) contour line was defined interactively in each GLI image with a graphics tablet. The central line between both contour lines was calculated with a skeleton algorithm, and equidistant profiles (spacing along the central line 200 μm) were automatically extracted from the GLI image. Each profile was oriented perpendicularly to the central line and extended from the outer to the inner contour line.

The profiles were standardized to a length of 101 GLI values (corresponding to a cortical depth of 100%) by resampling the data with linear interpolation in order to compensate for variations in cortical thickness. To quantify each profile's shape, five measures based on the laminar neuronal densities (mean y, mean x, standard deviation of x, skewness, kurtosis) were calculated for each profile and its first derivative and were combined into one feature vector **X**. The measures from the sample of n profiles per ROI were standardized to z-scores in order to assign equal weight to each of them. A mean feature vector $\overline{X_1}$ was calculated from a block of b_1 adjacent profiles ($8 \leq b_1 \leq 20$), and another mean feature vector $\overline{X_2}$ from a neighboring block of b_2 adjacent profiles ($8 \leq b_2 \leq 20$; $b_1 = b_2$). The Mahalanobis distance (Mahalanobis et al. 1949) $D^2 = \left(\overline{X_1} - \overline{X_2}\right)' C^{-1} \left(\overline{X_1} - \overline{X_2}\right)$ was calculated from the mean vectors $\overline{X_1}$ and $\overline{X_2}$ and the inverse of the pooled covariance matrix $\mathbf{C}^{-1}$. The D^2 values were plotted as a function of the positions of the profile blocks relative to the cortex. The resulting Mahalanobis distance function revealed maxima where the regions covered by profiles showed marked differences in their laminar patterns. Statistical significance was evaluated by a Hotelling's T^2 test

$$T^2 = \frac{D^2}{1/b_1 + 1/b_2}$$

($b_1 = b_2$, the number of profiles per block). The p values were Bonferroni-corrected for multiple comparisons. The number of profiles per block (b) defined the spatial resolution of this procedure, i.e., only cytoarchitectonic entities wider than the cortical sector covered by one block of profiles could be resolved with correct distance values. For this reason, significant maxima of the distance function D^2 were defined in the following way: (1) maxima with $p \geq 0.05$ were discarded; (2) if the distance between two adjacent maxima was equal to or smaller than one block of profiles, the maximum with the higher p value was discarded. The procedure converged after a second pass of step (2). The remaining maxima were termed "main maxima." For each ROI, this distance function was calculated for b=8 to b=20, and the positions of main maxima were plotted as a function of b. For further details see Schleicher et al. 1999. The positions of these main maxima were compared with the visible cytoarchitectonic pattern and marked on the coverslips of the cell-stained sections.

2.3 Three-Dimensional Reconstruction and Spatial Normalization of the Histological Volumes

Each mounted and cell-stained (i.e., each 60th coronal or 45th sagittal) histological section was digitized with a CCD camera (XC-75; Sony; image matrix 256×256 pixels; 8 bit gray value resolution). The histological volume of the brain was then reconstructed in 3-D from (1) the images of the paraffin blockface, (2) the digitized histological sections, and (3) the MR volume of the same brain (cf. Table 1) with linear and nonlinear transformations (Schormann et al. 1993, 1995, 1996, 1997; Schormann and Zilles 1997, 1998). Since the MR volume was obtained prior to histological processing, any artifacts (e.g., shrinkage of the brain due to dehydration in graded alcohols or compression of the sections due to cutting) could be eliminated in the 3-D reconstructed histological volume by matching it with the MR volume of the same brain. The statistically significant areal borders were marked on the histological sections. With an image analysis software package and an interactive voxel-painting program (KS 400; Zeiss) the extent of area 6 was labeled in the corresponding coronal or sagittal sections of the reconstructed volume. With this approach, a microstructurally defined representation of area 6 was obtained in the 3-D reconstructed histological volume of each brain.

Each 3-D reconstructed histological volume (with area 6) was spatially normalized to the reference brain (oriented in the Talairach coordinate system) of the computerized Human Brain Atlas (HBA; Roland et al. 1994; Roland and Zilles 1994) with an approach based on a new extended principal axes theory (PAT; Schormann et al. 1997) and a fast automated multiresolution full-multigrid movement model (FMG; Schormann and Zilles 1997, 1998).

The system of transformations was applied in a hierarchical manner from coarse to fine thereby combining low and high dimensional transformations. In a first step, the individual histological volumes were transformed to the reference brain with an extended PAT generalized to affine transformation parameters. This theory extends the classical PAT to non-rigid linear transformations that are at least one order of magnitude more accurate than the results of the classical PAT (Schormann and Zilles 1997). This technique is insensitive to noise or to symmetries of objects and does not require any interactive support for correlating corresponding landmarks (e.g., points, lines, or surfaces) in both volumes. Such a global or coarse registration accounts for differences in scaling (size), rotation (orientation), translation (position), and shearing that result from different eigensystems of both volumes (Schormann et al. 1993). The extended PAT precisely aligned objects on the basis of linear parameters; an important prerequisite for the nonlinear refinement by a fast automated full-multigrid movement technique (Schormann and Zilles 1997, 1998). The histological volumes were modeled as elastic material whereby the transformation to the reference brain was smooth, so that connected regions remained connected and global relationships between structures were maintained. In order to account for large deformations, it was necessary to extend the elastic model to a movement model whereby the histological volume was deformed iteratively by using the result of the previous iteration for the next step until a satisfactory match was achieved. High-dimensional transformations with up to 24 million degrees of freedom were used for the histological volumes in order to account for the different morphology between individual and reference brain whereby the transformations of the coarse and fine stage were determined without any interactive support. High-dimensional and linear transformations were combined and applied to each 3-D reconstructed histological volume with the representation of area 6.

The ten normalized histological volumes were superimposed in the 3-D space of the reference brain and a population map of area 6 was generated. This map shows the degree of interindividual microstructural variability of area 6 by exemplifying, for each voxel, how many brains have a representation of area 6 in this particular voxel.

3 Results

3.1 Cytoarchitectonic Features and Observer-Independent Analysis of the Border Between Area 6 and the Prefrontal Cortex

Cytoarchitectonic features of the non-primary motor cortex (Brodmann's area 6) that rostrally abuts on area 4 are an absent inner granular layer (layer IV; hence the term "agranular cortex") and fairly large and elongated pyramidal cells in lower layer III. Giant pyramidal (or Betz) cells, the most prominent feature of the primary motor cortex (Brodmann's area 4), are scattered throughout the caudal sector of area 6 close to the border with area 4. They are absent in the rostral part of area 6. At the

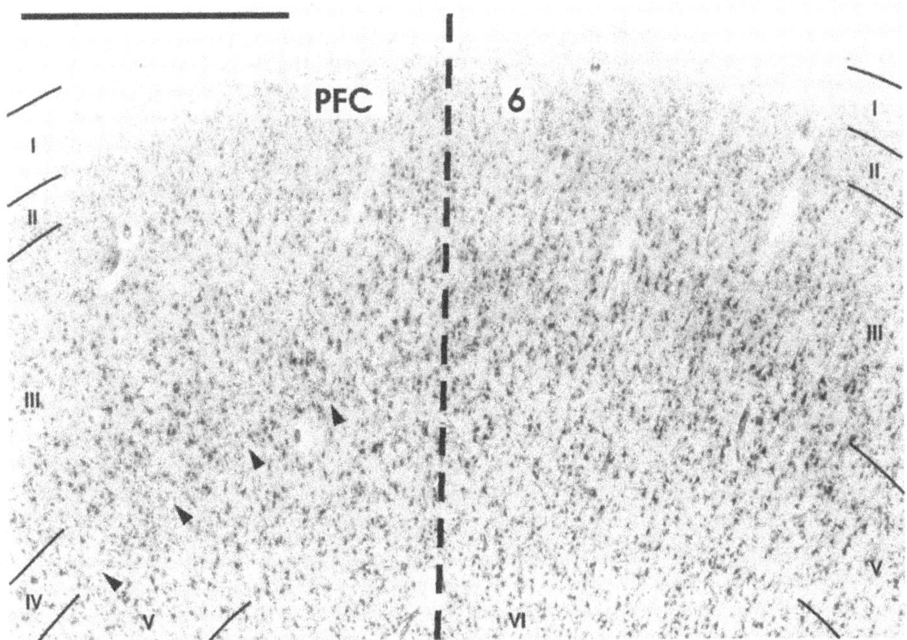

Fig. 1 Cytoarchitectonic features of the border between area 6 and the prefrontal cortex (*PFC*). Large and elongated pyramidal cells in lower layer *III* decrease in size and a discreet inner granular layer (layer *IV*; *arrowheads*) emerges. *Roman numerals* indicate cortical layers. Scale bar, 1 mm

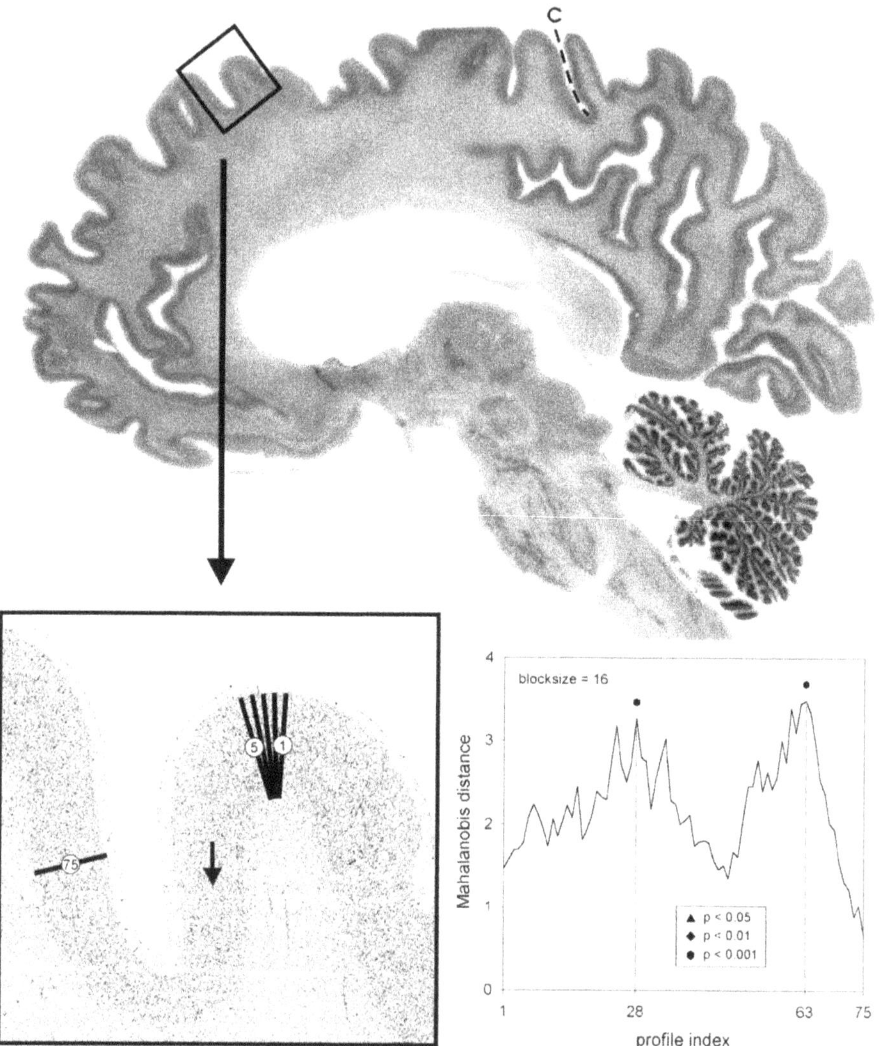

Fig. 2 *Top*: Sagittal whole brain section (no. 3226; brain no. 340/83) with ROI (*box*) that was used for calculation of the GLI (*C*, central sulcus). *Lower left*: GLI image of the ROI. Equidistant cell density profiles (spacing between adjacent profiles 200 μm) were extracted perpendicularly to the cortical layers from position no. 1–no. 75 (profiles at position no. 6–no. 74 not shown). *Lower right*: D^2 values of the Mahalanobis distance function (for b=16) plotted against profile positions. Main maxima and their levels of significance are marked

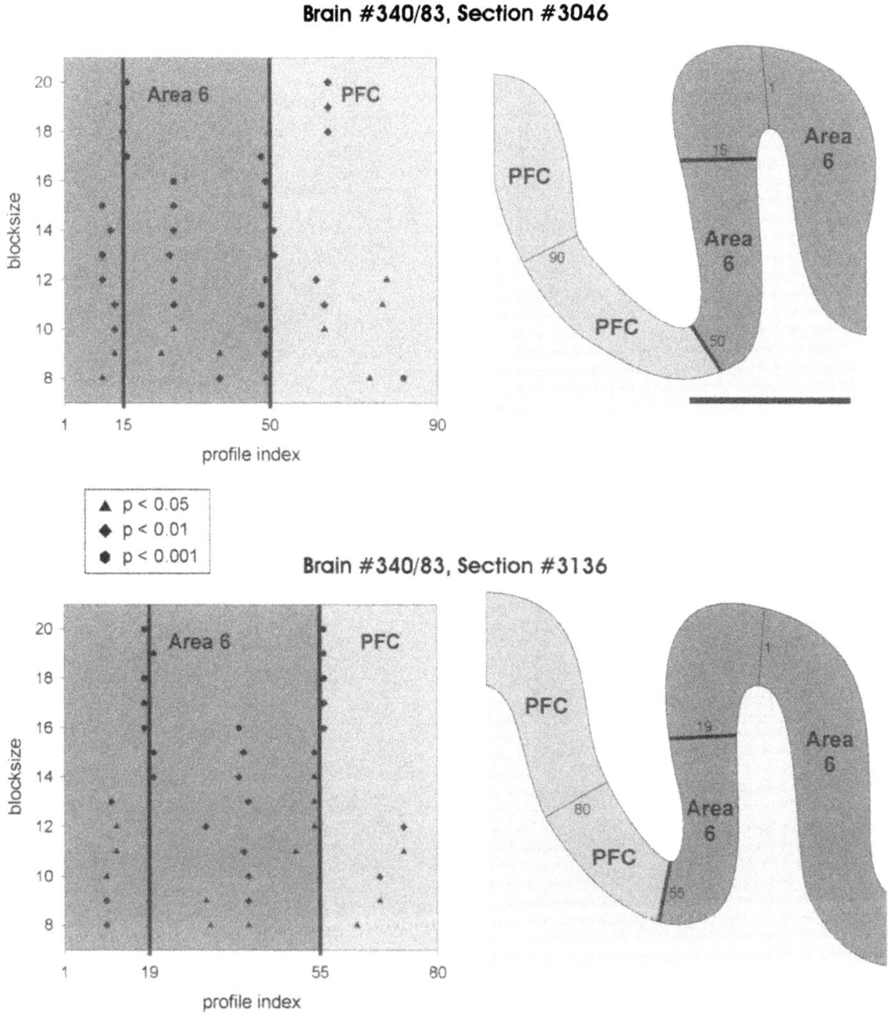

Brain #340/83, Section #3046

Brain #340/83, Section #3136

Fig. 3 Positions of main maxima (*abscissa*) and their levels of significance plotted as a function of *b* (*ordinate*) in sagittal sections no. 3046 (*top*) and no. 3136 (*bottom*; distance between sections 1.8 mm) from the same brain (no. 340/83). The border between area 6 (*dark gray*) and the PFC (*light gray*) is marked in the graphs and the schematic drawings of the ROIs. See text for further details. Scale bar, 5 mm

rostral border of area 6 with the prefrontal cortex (PFC) the large and elongated pyramids in lower layer III decrease in size (Fig. 1). The most important feature of the border between area 6 and the PFC is an emerging inner granular layer (layer IV; arrowheads in Fig. 1). In the caudal sector of the PFC, layer IV is discreet and barely detectable (dysgranular cortex). Further rostral, layer IV gradually increases in width and cell density. Hence, the features of the "typical" granular cortex emerge gradually rather than abruptly.

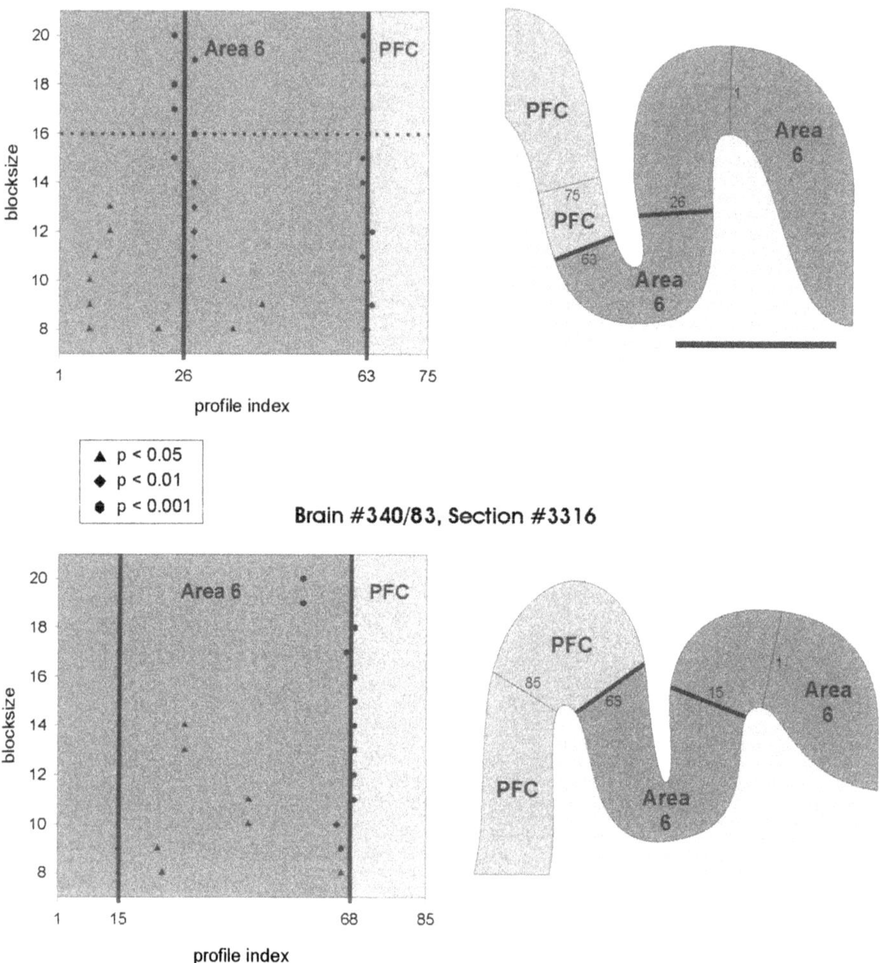

Fig. 4 Positions of main maxima (*abscissa*) and their levels of significance plotted as a function of b (*ordinate*) in sagittal sections no. 3226 (*top*) and no. 3316 (*bottom*; distance between sections 1.8 mm) from the same brain (no. 340/83). *Dotted horizontal line* (b=16) corresponds to the plot in Fig. 2. The border between area 6 (*dark gray*) and the PFC (*light gray*) is marked in the graphs and the schematic drawings of the ROIs. See text for further details. Scale bar, 5 mm

Since it may sometimes be difficult to subjectively detect where layer IV discreetly emerges, this region was studied in more detail with an observer-independent approach.

The results are shown in two brains, sectioned in a sagittal plane: brain no. 340/83 (Figs. 2, 3, 4) and brain no. 646/79 (Figs. 5, 6, 7). Brain no. 340/83 is depicted in Fig. 2. The sagittal whole brain section (no. 3226) intersects the right hemisphere close to the midline; the rostral and caudal end of the cingulate sulcus and the brain stem are visible. Brain no. 646/79 is depicted in Fig. 5. The sagittal whole brain sec-

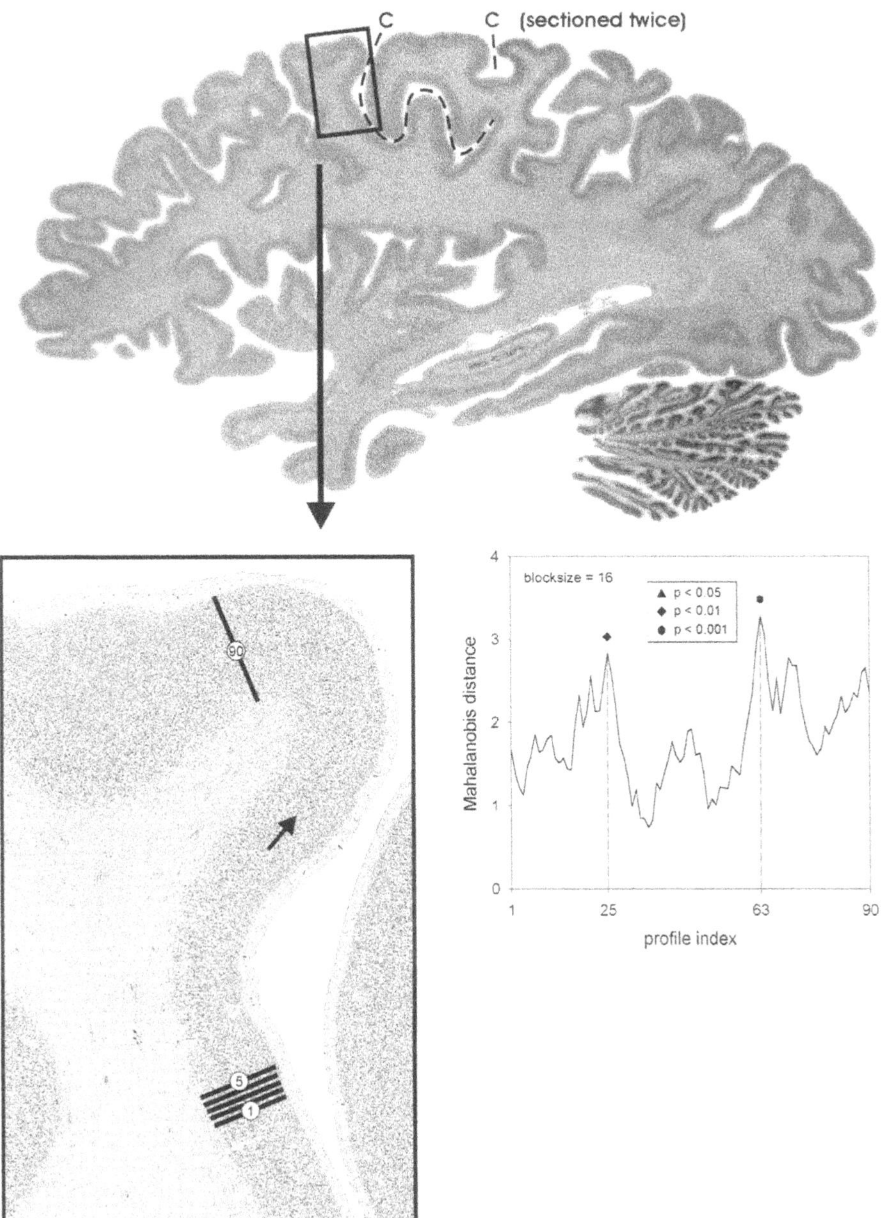

Fig. 5 *Top*: Sagittal whole brain section (no. 1441; brain no. 646/79) with ROI (*box*) that was used for calculation of the GLI [*C*, central sulcus (sectioned twice)]. *Lower left*: GLI image of the ROI. Equidistant cell density profiles (spacing between adjacent profiles 200 μm) were extracted perpendicularly to the cortical layers from position no. 1–no. 90 (profiles at position no. 6–no. 89 not shown). *Lower right*: D^2 values of the Mahalanobis distance function (for b=16) plotted against profile positions. Main maxima and their levels of significance are marked

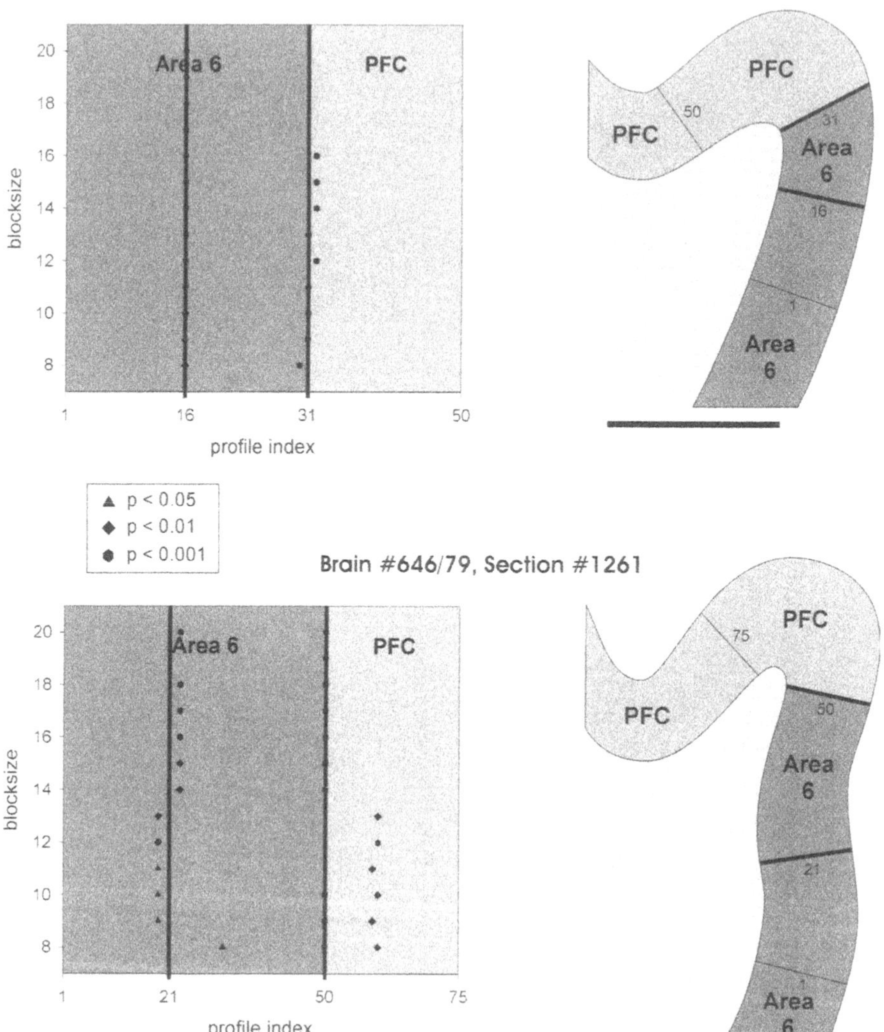

Fig. 6 Positions of main maxima (*abscissa*) and their levels of significance plotted as a function of *b* (*ordinate*) in sagittal sections no. 1171 (*top*) and no. 1261 (*bottom*; distance between sections 1.8 mm) from the same brain (no. 646/79). The border between area 6 (*dark gray*) and the PFC (*light gray*) is marked in the graphs and the schematic drawings of the ROIs. See text for further details. Scale bar, 5 mm

tion (no. 1441) intersects the left hemisphere further laterally; the gray matter of the insula and the hippocampus and lateral ventricle in the temporal lobe are visible. Boxes mark the ROIs that were used for calculation of the GLI and subsequent extraction of profiles. The GLI image of each ROI is shown at a higher magnification in Figs. 2 (lower left) and 5 (lower left). From these GLI images, equidistant cell density

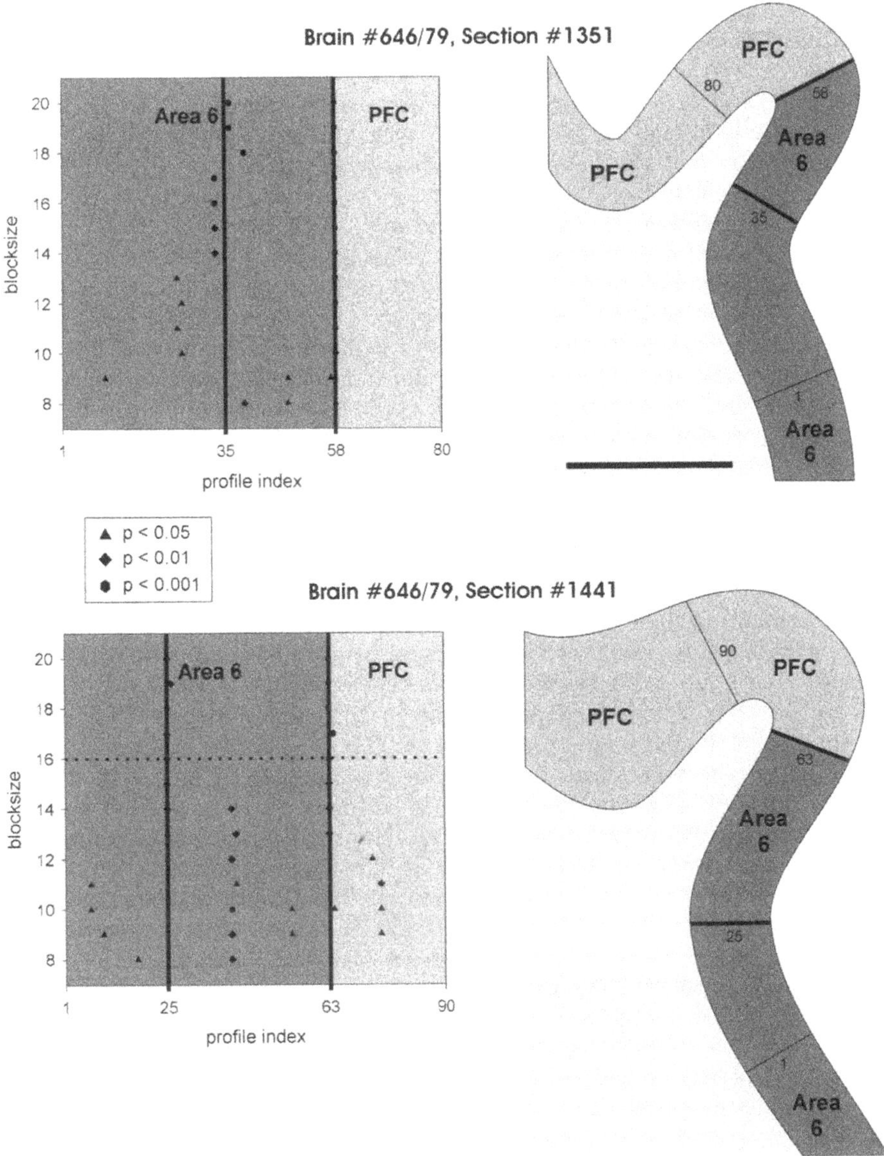

Fig. 7 Positions of main maxima (*abscissa*) and their levels of significance plotted as a function of *b* (*ordinate*) in sagittal sections no. 1351 (*top*) and no. 1441 (*bottom*; distance between sections 1.8 mm) from the same brain (no. 646/79). *Dotted horizontal line* (b=16) corresponds to the plot in Fig. 5. The border between area 6 (*dark gray*) and the PFC (*light gray*) is marked in the graphs and the schematic drawings of the ROIs. See text for further details. Scale bar, 5 mm

profiles (125 μm wide; spacing between adjacent profiles 200 μm) were extracted perpendicularly to the cortical layers from the border between layers I and II to the border between layer VI and the white matter. The positions of the first five profiles (no. 1–no. 5) and the last profile (no. 75 in Fig. 2 and no. 90 in Fig. 5) are marked. In Figs. 2 (lower right) and 5 (lower right), D^2 values (for b=16) are plotted against the positions of the profiles in the cortex. Two statistically significant main maxima were detected in each ROI: in Fig. 2 at positions no. 28 ($P<0.001$) and no. 63 ($P<0.001$) and in Fig. 5 at positions no. 25 ($P<0.01$) and no. 63 ($P<0.001$).

For the same set of profiles (no. 1–no. 75 in Fig. 2 and no. 1–no. 90 in Fig. 5), main maxima were determined for b=8 through b=20 and their positions plotted as a function of b. The results are shown in Fig. 4 (top) and Fig. 7 (bottom), respectively. Only the main maxima positions and their levels of significance are indicated; the dotted horizontal line (b= 16) corresponds to the plot shown in Figs. 2 and 5, respectively. With higher numbers of profiles per block (i.e., higher values of b), the degree of smoothing increases and, as a consequence, the number of main maxima decreases. In each figure, however, the positions of two maxima are remarkably stable. Across a wide range of b, the two maxima at positions no. 26 and no. 63 in Fig. 4 (top) and no. 25 and no. 63 in Fig. 7 (bottom) can be reproduced at similar positions in the cortex [positions are marked in the schematic drawings of the cortex in Fig. 4 (top) and Fig. 7 (bottom)].

The positions of these main maxima can be reproduced at comparable positions across a sequence of several sections (distance between adjacent sections 1.8 mm) from the same brain. In the brain no. 340/83, the maximum at position no. 15 in section no. 3046 (Fig. 3, top) corresponds to the maxima at positions no. 19 in section no. 3136 (Fig. 3, bottom), no. 26 in section no. 3226 (Fig. 4, top), and no. 15 in section no. 3316 (Fig. 4, bottom). The maximum at position no. 50 in section no. 3046 (Fig. 3, top) corresponds to the maxima at positions no. 55 in section no. 3136 (Fig. 3, bottom), no. 63 in section no. 3226 (Fig. 4, top), and no. 68 in section no. 3316 (Fig. 4, bottom). In the brain no. 646/79, the maxima at positions no. 16 and no. 31 in section no. 1171 (Fig. 6, top) correspond to the maxima no. 21 and no. 50 in section no. 1261 (Fig. 6, bottom), no. 35 and no. 58 in section no. 1351 (Fig. 7, top), and no. 25 and no. 63 in section no. 1441 (Fig. 7, bottom).

In the brain no. 340/83, the maxima at position no. 50 in section no. 3046 (Fig. 3, top) and at corresponding positions in the other sections of the sequence coincide with a discreetly emerging layer IV and thus indicate the border between area 6 and the PFC. The maxima at position no. 15 in section no. 3046 (Fig. 3, top) and at corresponding positions in the other sections of the sequence do not coincide with a visible change in the cytoarchitectonic pattern. In the brain no. 646/79, the maxima at position no. 31 in section no. 1171 (Fig. 6, top) and at corresponding positions in the other sections of the sequence also coincide with a discreetly emerging layer IV and thus indicate the border between area 6 and the PFC. The maxima at position no. 16 in section no. 1171 (Fig. 6, top) and at corresponding positions in the other sections of the sequence do not coincide with a visible change in the cytoarchitectonic pattern.

3.2
Cytoarchitectonic Features and Observer-Independent Analysis of the Border Between Area 6 and the Primary Motor Cortex

The border between area 6 and the primary motor cortex (Brodmann's area 4) will be described only briefly here, since the caudal border of area 6 (or the rostral border of area 4) has extensively been described in a previous observer-independent mapping study of area 4 (Geyer et al. 1996). Area 6 is characterized by large and elongated pyramidal cells in lower layer III; at the border between area 6 and 4 they decrease in size. Giant pyramidal (Betz) cells are scattered throughout the caudal sector of area 6; at the border between area 6 and 4 they abruptly increase in size and density (cf. Fig. 23 in the discussion of the structural organization of the isocortical motor system in humans).

The border between area 6 and 4 can also be delineated in an observer-independent way. Numerical data can be found in Geyer et al. 1996.

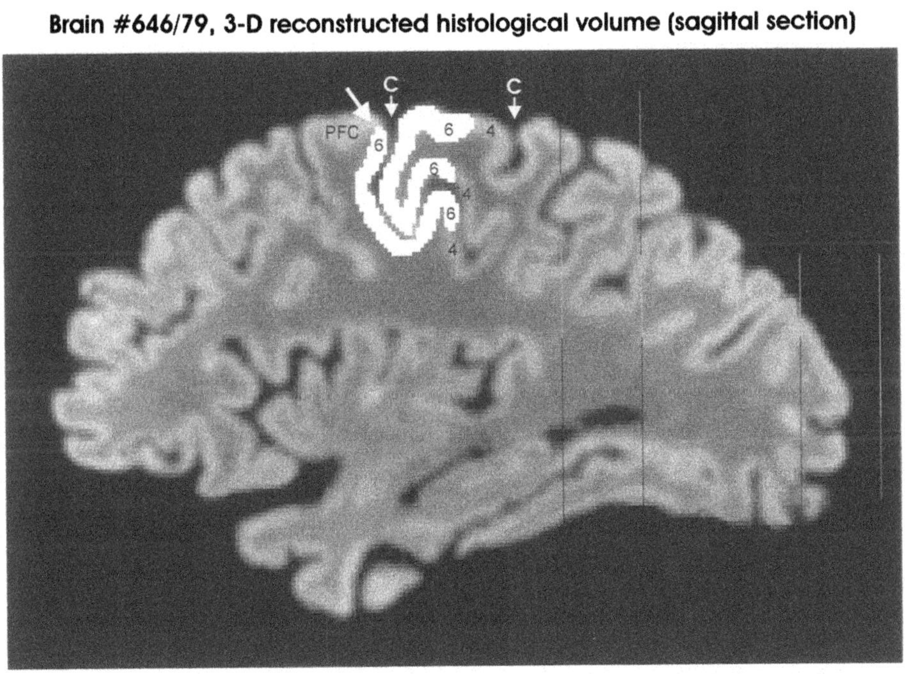

Brain #646/79, 3-D reconstructed histological volume (sagittal section)

Fig. 8 Sagittal section through the reconstructed histological volume (brain no. 646/79) that corresponds to section no. 1441 of the same brain (cf. Fig. 5). Area 6 is marked in *white*; the topography of the border between area 6 and the PFC (*arrow*) corresponds to the schematic drawing in Fig. 7 (*bottom*). Numeral *4* is Brodmann's area 4; *C*, central sulcus (sectioned twice)

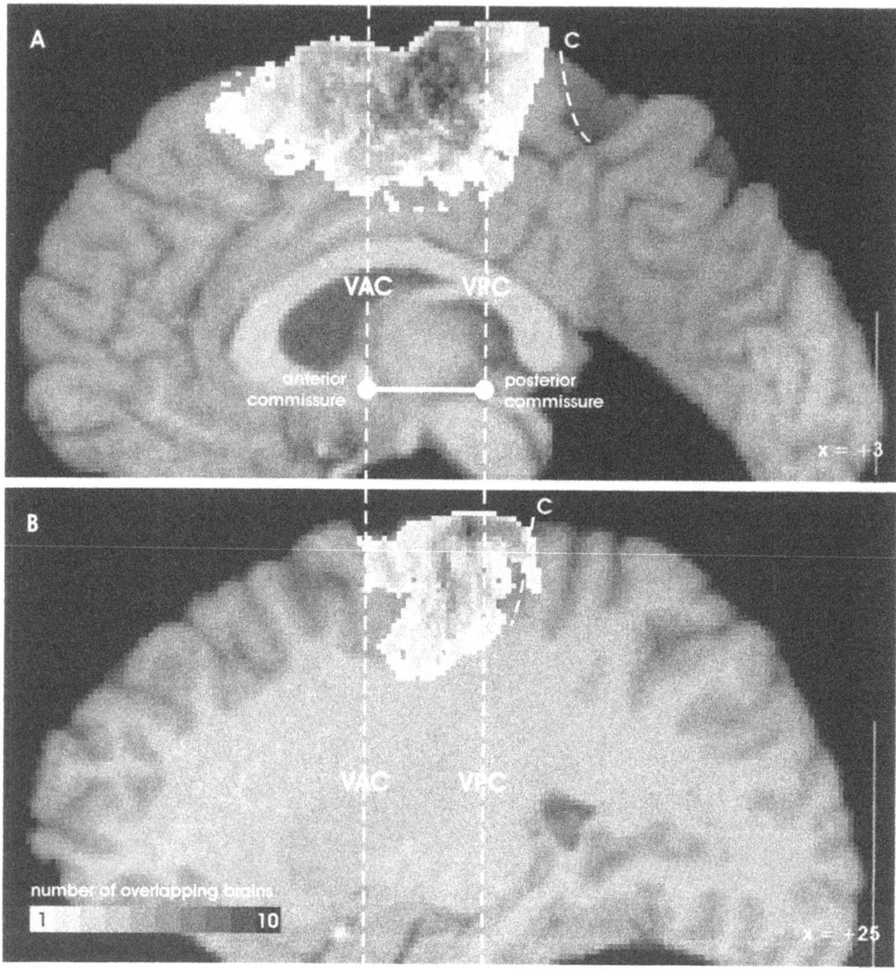

Fig. 9A–D Cytoarchitectonic population map of area 6. The sagittal section in **A** intersects the right hemisphere 3 mm lateral (x=+3) to the midline, section in **B** 25 mm lateral (x=+25) thereto, section in **C** 33mm lateral (x=+33) thereto, and section in **D** 41mm lateral (x=+41) thereto. In each voxel the number of brains ($1 \leq n \leq 10$) that have a representation of area 6 is coded as a *gray value* (cf. bar). Reference brain of the computerized atlas is shown in the *background*. *C*, central sulcus; *VAC*, VAC line (traverses the anterior commissure vertical to a line passing through the anterior and posterior commissure); *VPC*, VPC line (traverses the posterior commissure vertical to a line passing through both commissures)

3.3 Reconstruction of the Histological Volumes in 3-D, Spatial Normalization, and Generation of the Population Map of Area 6

With an interactive voxel-painting program, the extent of area 6 was labeled in the corresponding sections of the reconstructed volume. Thus, the microstructurally de-

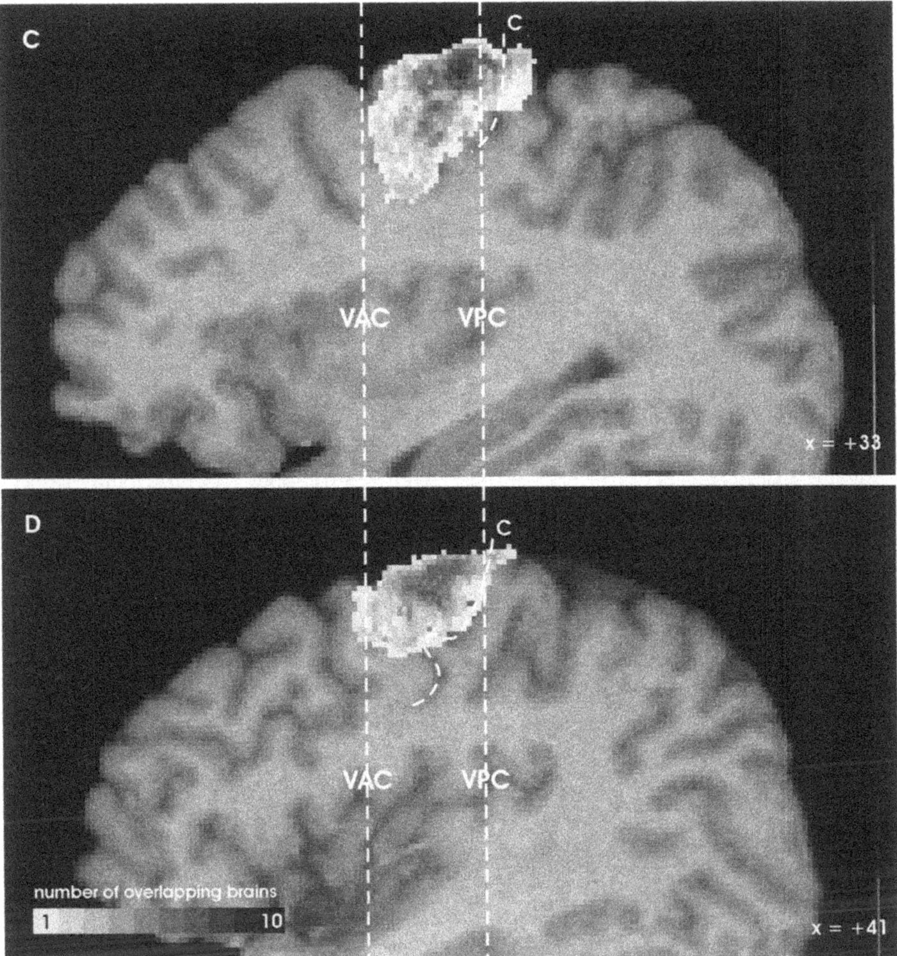

Fig. 9 C–D

fined representation of area 6 was transferred from the histological sections to the 3-D-reconstructed histological volume of the same brain. Fig. 8 shows a sagittal section through the reconstructed volume (brain no. 646/79) that corresponds to section no. 1441 of the same brain (cf. Fig. 5). Area 6 is marked in white; the topography of the border between area 6 and the PFC (arrow) corresponds to the schematic drawing in Fig. 7 (bottom).

Each histological volume (with the representation of area 6) was then spatially normalized to the reference brain of the computerized atlas. The ten normalized volumes were superimposed in the 3-D space of the reference brain and a population map of area 6 was generated (Fig. 9). This map quantifies the degree of interindividual microstructural variability of area 6 by exemplifying, for each voxel, how many brains have a representation of area 6 in this particular voxel. This representation is coded from light gray (area 6 present in 1 of 10 brains in this voxel), to black (present in all 10 brains).

On the mesial cortical surface (Fig. 9A), the border between area 6 and the PFC lies in most brains rostral to the anterior commissure [or vertical anterior commissure (VAC) line that traverses the anterior commissure vertically to a line passing through the anterior and posterior commissure]. Interindividual variability in the topography of the area 6/PFC border is high. Maximum overlap (area 6 in $n \geq 6$ brains) can be found caudal to the VAC line; rostral to it, the degree of overlap decreases gradually. The border between area 6 and the primary motor cortex (area 4) coincides approximately with the posterior commissure [or vertical posterior commissure (VPC) line that traverses the posterior commissure vertically to a line passing through both commissures]. Individual variability in the topography of the area 6/area 4 border is considerably lower. The degree of overlap of area 6 sharply drops at the VPC line; caudal to it, the degree of overlap is low (area 6 in $n \leq 3$ brains).

On the lateral convexity (Fig. 9B–D), the border between area 6 and the PFC recedes in a caudal direction. Close to the Sylvian fissure, area 6 occupies only the superficial part of the precentral gyrus (Fig. 9D).

There is no macroanatomical landmark in humans that separates area 6 from the prefrontal cortex. The question whether a motor task engages only the "motor domain" of the supplementary motor and/or lateral premotor cortex, or in addition the "cognitive domain" of the prefrontal cortex can only be answered by superimposing the functional activation map upon the microstructural population map of area 6.

4 Discussion

4.1 Observer-Independent Delineation of Area 6

In contrast to earlier attempts to map the human agranular frontal cortex (Bailey and von Bonin 1951; Braak 1980; Brodmann 1909; Campbell 1905; Sarkissov et al. 1955; Smith 1907; Vogt and Vogt 1919; von Economo and Koskinas 1925) this approach defines cytoarchitectonic borders in an objective way. This obviates subjective evaluation of differences in the cytoarchitectonic pattern (and all problems related to it) that has so far been the only way to structurally define cortical areas. Since the algorithm detects only "borders" (i.e., significant changes in the cytoarchitectonic pattern), it is then necessary to interpret and classify the entities surrounded by these borders, e.g., by comparing the cytoarchitectonic features of these entities with verbal or pictorial descriptions published in the literature.

The observer-independent technique is based on the assumption that cytoarchitectonic entities are characterized by a specific laminar pattern that differs from that of neighboring entities. Density profiles extracted perpendicularly to the pial surface across cortical layers II–VI quantify this pattern. Maximal differences in profile shapes can be expected when two groups of profiles sampled from two different entities are compared. Single profiles (although 125 μm wide) are affected by local structural inhomogeneities, e.g., cell clusters or blood vessels. To smooth the signal and improve the signal-to-noise ratio of the distance function, two adjacent *groups* of profiles were compared instead of two adjacent single profiles (see also Schleicher et al. 1999).

Nonetheless, several factors are potential sources of false positive maxima. Variations in staining intensity of cell bodies and/or background are irrelevant because the profiles have been extracted not from digitized images of the cell-stained sections but instead from GLI images in which cell bodies have been segmented from the background by adaptive thresholding (Schleicher and Zilles 1990). Artifacts due to sectioning of the tissue (e.g., tears) are a problem when analyzing whole brain sections. However, the number of profiles per block (b) limits the spatial resolution of the algorithm (from b=8×200 μm = 1.6 mm to b=20×200 μm = 4 mm). Only entities wider than a cortical sector covered by one block of profiles can be resolved with correct distance values. Furthermore, artifacts are usually distributed randomly throughout the cortex, and this should become evident when comparing a series of consecutive sections. Cortical arteries and veins rarely exceed 200 μm in diameter and they are oriented more or less perpendicularly to the cortical layers. Smaller vessels arise from the principal trunk, and spread out at different angles but their length parallel to the cortical layers does not reach the threshold of spatial resolution (Duvernoy et al. 1981). In gyrencephalic brains, inward and outward folds of the cortex cause distortions of the cytoarchitectonic pattern (cf., e.g., Van Essen 1997). In order to assess to what extent cortical folding causes false-positive borders, we generated a model of a cortical region with layers and added step-like changes in GLI values oriented perpendicularly to the layers (i.e., "borders") and a low frequency sine-shaped thinning of layer II (i.e., a distortion in the thickness of layers II and III induced by cortical folding; Schleicher et al. 1999). The distance function detected all "borders" with sharp and significant maxima but not the sine-shaped distortion (Schleicher et al. 1999). In addition, Figs. 2–7 of this study show that those positions where the curvature of the cortex is maximal (e.g., fundi of sulci) do not coincide with positions of significant maxima.

The next step is to interpret and classify the cytoarchitectonic entities by comparing their features with descriptions published in the literature.

The most famous cytoarchitectonic parcellation is Brodmann's (1909) map. Unfortunately, Brodmann's publications (1903, 1908, 1909) contain almost no verbal descriptions and no photomicrographs of each area's cytoarchitectonic features. Instead, he only briefly comments on their topography. The map shows each area's location and extent only on the exposed cortical surface. The sulci are not opened up and no information is available on the precise areal topography within each sulcus. In the rostral wall of the central sulcus Brodmann defines "area 4" or "area gigantopyramidalis". In his article "Beiträge zur histologischen Lokalisation der Großhirnrinde. Erste Mitteilung: Die Regio Rolandica," published in 1903, Brodmann describes the cytoarchitectonic features of area 4 in greater detail: unusually wide gray matter, poor lamination, low cell density, absence of an inner granular layer (layer IV), large pyramidal cells (Betz cells) in layer V, and a diffuse border between layer VI and the white matter. Area 4 occupies a wedge-shaped cortical region on the precentral gyrus and the adjoining part of the paracentral lobule. On the mesial cortical surface, area 4 lies in the middle third of the paracentral lobule. On the lateral convexity, area 4 occupies the precentral gyrus close to the midline across its entire rostrocaudal extent and abuts the base of the superior frontal gyrus. Further laterally (toward the Sylvian fissure), area 4 recedes in a caudal direction and eventually disappears in the depth of the central sulcus (Brodmann 1909). Rostral to area 4, Brodmann defines "area 6" or "area frontalis agranularis". He only briefly comments on its cytoarchitecture. Area 6 is related to area 4 due to its agranular nature, and, together with area 4, should be considered as one group of areas that can be separated from the granular frontal areas (Brodmann 1909). Location and extent of area 6 are described in greater detail. Area 6 occupies a band-like cortical region (its rostrocaudal extent narrowing in a mediolateral direction) and extends from the callosomarginal sulcus on the mesial surface across the frontal lobe to the upper margin of the Sylvian fissure. On the mesial surface, area 6 lies in the rostral part of the paracentral lobule and extends to the superior frontal gyrus. On the lateral convexity, area 6 lies on the superior and middle frontal gyrus and on that part of the precentral gyrus not occupied by area 4 (Brodmann 1909).

In contrast to Brodmann, von Economo and Koskinas (1925) meticulously describe each area's topography and cytoarchitecture, and they also include photomicrographs. On the crown and in the caudal wall of the precentral gyrus (close to the Sylvian fissure only in the caudal wall) they define "area praecentralis (area FA)" with low cell density and a missing layer IV. A subregion of area FA, characterized by giant pyramidal cells in layer V, is termed "area praecentralis gigantopyramidalis (area FAγ)". Rostral to area FA, they delineate "area frontalis agranularis (area FB)". Area FB has the shape of a triangle, its base lying on the mesial cortical surface and its apex pointing on the lateral cortical convexity toward the rostral end of the Sylvian fissure. Cytoarchitectonic features of area FB are a missing layer IV and large and elongated pyramids in lower layer III sometimes arranged in parallel rows like a phalanx. Further rostral, "area frontalis intermedia (area FC)" occupies the superior and middle, "area frontalis intermedio-agranularis magnocellularis in Campo Broca (area FCBm)," the inferior frontal gyrus. Despite the misleading nomenclature of area FCBm ("intermedio-agranularis"), the photomicrographs of both area FC and FCBm show an incipient layer IV. Hence, both areas are "dysgranular." Rostral to areas FC and FCBm lies "area frontalis granularis (area FD)" with a fully developed layer IV.

The descriptions published by Brodmann (1909) and von Economo and Koskinas (1925) show that the cytoarchitectonic entity delineated in this study corresponds to Brodmann's area 6 or von Economo and Koskinas's area FB. Since Brodmann's terminology is much more common in the neuroscience literature, I adopted his nomenclature and termed the region "area 6". In Brodmann's map, several areas (8, 9, and 44) rostrally abut area 6. Hence, for the region rostral to area 6, I adopted a more general name and termed it "prefrontal cortex (PFC)."

An unresolved issue is the functional meaning of those borders within area 6 that have been detected with the observer-independent algorithm but which do not coincide with visible changes in the cytoarchitectonic pattern. As will be shown later in the discussion, it is now known that the primate cortical motor system is made up of a multiplicity of different areas. Over the last years, their structural and functional properties have been elucidated in more and more detail in the macaque monkey. There is also increasing evidence that the cortical motor system of macaques and humans is organized in a similar way (see further below). These "additional" borders could be the structural equivalents of a similarly complex functional organization of this cortical region in humans. It is conceivable that the cytoarchitectonic borders are so subtle that they can be resolved only with an observer-independent technique.

4.2 Spatial Normalization of Area 6

As a next step, each histological volume (with the representation of area 6) was spatially normalized to the reference brain of the computerized atlas. The reference brain was selected among the MR scans of 21 brains from 20–30-year-old healthy, right-handed, male volunteers, and was defined as that individual brain that deviates the least in shape from the 20 other brains (Roland et al. 1994). This brain (oriented in Talairach space) serves as a common basis to integrate microstructural and functional data.

For spatial normalization, a new and mathematically more powerful warping technique has been used in this study that does not require any user interaction. Interactive approaches to normalize brains cannot be reproduced among different investigators and the quality of fit depends on each investigator's experience. The approach is based on a new extended principal-axes theory (Schormann et al. 1997) and a fast automated multiresolution full-multigrid movement model (Schormann and Zilles 1997, 1998). The power and accuracy of this technique in comparison with other warping approaches has recently been tested (Crivello et al. 1999, 2002) with 20 MR brain scans that were spatially normalized to the reference brain with four different warping techniques: (1) a simple nine parameter model, (2) the nonlinear procedure implemented in the statistical parametric mapping (SPM) package (Ashburner and Friston 1997), (3) the 5th order (168 parameters) nonlinear warping algorithm included in the automated image registration (AIR) tool (Woods et al. 1998), and (4) the multiresolution full-multigrid movement model (FMG; used in this study). The adjusted mean MR volume across the 20 subjects was computed for each procedure, and difference volumes (mean volume$_{SPM}$−mean volume$_{FMG}$ and mean volume$_{SPM}$−mean volume$_{AIR}$) were generated. The SPM and AIR algorithms performed similarly whereas the FMG approach dramatically improved the spatial accuracy with which each individual brain was adapted to the reference brain (Crivello et al. 1999, 2002). With this new FMG approach, the residual macroanatomical variability is dramatically reduced (although not completely eliminated). The outer contour and the ventricular surface of a mean brain generated from ten histological volumes that have been normalized with the FMG technique are sharp, but sulci and subcortical nuclei are slightly blurred (Geyer et al. 2000b). It can therefore be assumed that the "interindividual variability" of area 6 as shown in the population map is due to minimal residual macroanatomical variability, but the rest is indeed microstructural in nature.

4.3 The Isocortical Motor System in Macaques

Until the second half of the twentieth century, the concept prevailed that the primate cortical motor system is made up of three functional entities: the primary motor cortex (area 4) in the caudal wall and on the crown of the precentral gyrus, the supplementary motor area further rostral on the mesial surface of the frontal lobe (mesial part of area 6), and the premotor cortex on the lateral surface of the frontal lobe (lateral part of area 6).

Over the last years, new and more powerful anatomical and functional techniques have shown that this view is no longer adequate: instead of three functional entities, the primate cortical motor system is made up of many structural and functional fields, each of which processes different aspects of motor behavior. Similar to the motor cortex, a multiplicity of areas has been defined in the posterior parietal cortex (including the intraparietal sulcus), each of which is involved in the analysis of particular aspects of visual and/or somatosensory information. Parietal and frontal motor areas are connected in a specific pattern thus forming several parieto–frontal circuits. These circuits work in parallel and transform different aspects of sensory information into appropriate motor commands.

This part of the discussion gives an overview of the anatomical and functional organization of the cortical motor system in macaques and discusses possible mechanisms that explain parietal and frontal area interaction when goal-directed movements are performed. I confine myself to the six-layered isocortex; the proisocortical cingulate motor areas are not described. The structural basis in these latter areas is

Table 3 Synopsis of different parcellation schemes of the macaque agranular frontal cortex (modified from Schieber 1999)

Generic nomenclature	Generic abbreviation	Brodmann (1909)	Vogt and Vogt (1919)	von Bonin and Bailey (1947)	Barbas and Pandya (1987)	Matelli et al. (1985, 1991)
Primary motor cortex	M1	4	4a, 4b, 4c	FA	4	F1
Dorsocaudal part of premotor cortex	PMdc	6	6aα	FB	6DC	F2
Supplementary motor area proper	SMA proper					F3
Ventrocaudal part of premotor cortex	PMvc			FBA	6Va	F4
Ventrorostral part of premotor cortex	PMvr		6bα, 6bβ	FCBm	6Vb	F5
Pre-supplementary motor area	Pre-SMA		6aβ	FC	MII	F6
Dorsorostral part of premotor cortex	PMdr				6DR	F7

still poorly established and the maps proposed by various authors in macaques differ from each other in fundamental ways. I conclude the discussion with a current overview of the structural and functional organization of the isocortical motor system in humans.

4.3.1 The Structural Framework: Microanatomical Subdivision of the Agranular Frontal Isocortex

Various architectonical maps of the agranular cortex (i.e., layer IV or internal granular layer is missing) of the macaque monkey have been published during the twentieth century (Barbas and Pandya 1987; Brodmann 1909; Matelli et al. 1985, 1991; Vogt and Vogt 1919; von Bonin and Bailey 1947). Most investigators agree that the primary motor cortex is homogeneous, whereas the rostrally adjoining agranular cortex can be subdivided into three groups of areas: the supplementary motor areas "SMA proper" and "pre-SMA" on the mesial cortical surface, the dorsal part of the premotor cortex (PMd) on the dorsolateral convexity, and the ventral part of the premotor cortex (PMv) on the ventrolateral convexity. A synopsis of the various parcellation schemes (Table 3) reveals two aspects.

First, the maps have become more and more complex over time. Brodmann (1909) defined two areas, whereas seven areas can be found in the map of Matelli et al. (1985, 1991)—a consequence of more sensitive techniques that have been developed in recent years and that have revealed an increasing number of structural details.

Second, the maps vary in terms of size, extent, and topography of areas—a consequence of each investigator's subjective criteria for defining borders between areas.

For example, the most widely used criterion for defining the border between Brodmann's area 4 and 6 is the change in number and density of giant pyramidal (Betz) cells in layer V. However, these cells do not stop abruptly at the area 4/area 6 border. Instead, their density decreases gradually and there is considerable interindividual variability. If this criterion alone is used, the definition of the area 4/area 6 border is more or less arbitrary. This single criterion was used by Barbas and Pandya (1987) for delineating area 6DC (where giant pyramids are scattered). The mesial part of area 6DC, however, is the leg representation of the SMA proper (see below; Luppino et al. 1991, 1993; Mitz and Wise 1987) and cannot be considered a functionally independent field. On the other hand, when areas are defined on the basis of a combination of (1) different criteria and (2) different technical approaches, their location and extent becomes more reliable and they indeed reflect structurally and functionally different entities. The "F" nomenclature of Matelli et al. (cf. Table 3, last column) that is used throughout this book was introduced in 1985 based on regional differences in cytochrome oxidase histochemistry. The terminology was adopted for two reasons. First, the location of the enzymatic areas was very similar to the cytoarchitectonic areas of von Bonin and Bailey (1947), hence a terminology similar to their nomenclature ["F" for frontal cortex; instead of capital letters for each area (FA, FB, etc.), as introduced by von Bonin and Bailey, the areas were numbered in a caudo-rostral sequence]. Second, since Matelli et al. wanted to avoid functional terms such as M1, supplementary motor area, etc., they adopted a more "neutral" nomenclature. This parcellation has subsequently been confirmed with other techniques, such as determination of cytoarchitecture (Matelli et al. 1991), receptor autoradiographic mapping of binding sites of classical neurotransmitters (Geyer et al. 1998a; Zilles et al. 1995), and, most recently, immunohistochemical staining of neurofilament proteins with monoclonal antibody SMI-32 (Gabernet et al. 1999; Geyer et al. 1998b, 2000c; Matelli et al. 1996; Petrides et al. 2000). For example, areas F3 and F6 according to this nomenclature indeed reflect two different functional entities (SMA proper and pre-SMA, respectively; Rizzolatti et al. 1996c; Tanji 1994).

According to the nomenclature of Matelli et al., areas F1 to F7 represent the agranular frontal isocortex (Fig. 10A, C). Its caudal border (with the primary somatosensory cortex) lies close to the fundus of the central sulcus and its mesial border (with the cingulate cortex) in the upper bank of the cingulate sulcus. Its rostral border (with the prefrontal cortex) lies close to the fundus of the arcuate sulcus and looks like a horizontally mirrored "c". Only its lateral border on the cortical convexity does not coincide with a macroanatomical landmark. It runs approximately from the lateral end of the inferior arcuate sulcus to the lateral end of the central sulcus. Area F1 corresponds to the primary motor cortex. Areas F2 to F7 lie further rostrally: areas F3 and F6 on the mesial cortical surface, areas F2 and F7 on the dorsolateral cortical convexity (between the midline and the level of the spur of the arcuate sulcus), and areas F4 and F5 on the ventrolateral convexity (between the spur of the arcuate sulcus and the lateral border of the agranular frontal cortex).

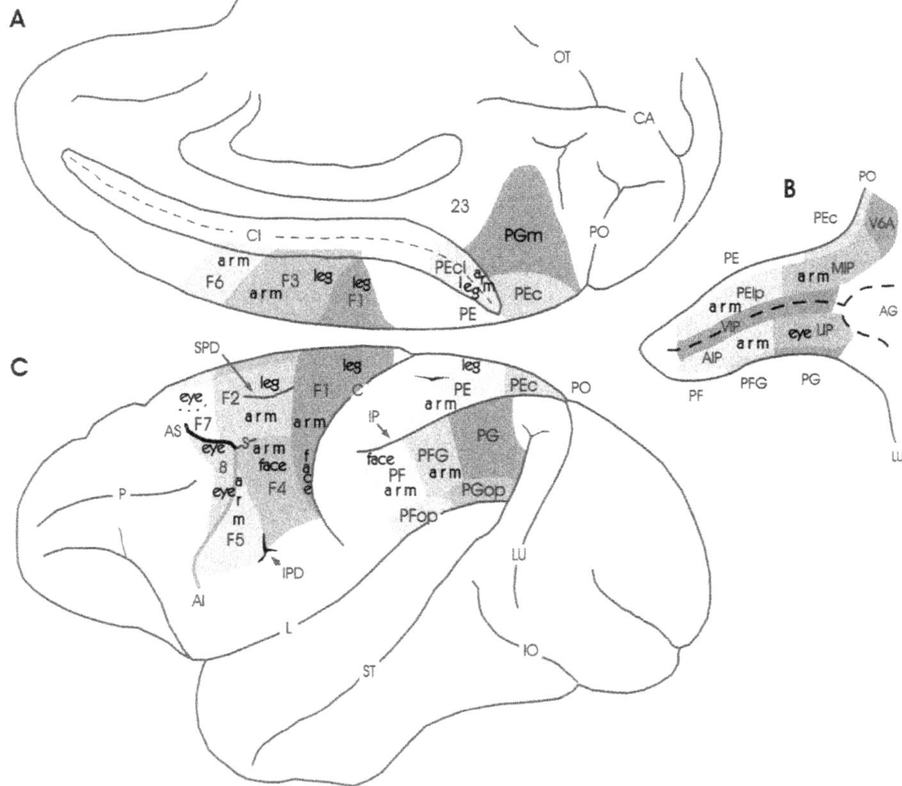

Fig. 10A–C Mesial (**A**) and lateral (**C**) view of the macaque brain with the areas of the agranular frontal cortex according to Matelli et al. (1985, 1991), and the posterior parietal cortex according to Pandya and Seltzer (1982)—plus the somatotopic representations of the body. The areas in the depth the intraparietal sulcus (*IP*) are defined according to electrophysiological data (see text) and are depicted in an unfolded view of the sulcus (**B**; *upper half*, medial bank; *lower half*, lateral bank; *dashed line*, fundus of the sulcus). See Fig. 25 for gray values of the superior (*AS*) and inferior (*AI*) arcuate sulcus. Numeral *8* is Brodmann's area 8 (frontal eye field); *AG*, annectant gyrus; *C*, central sulcus; *CA*, calcarine sulcus; *CI*, cingulate sulcus; *IO*, inferior occipital sulcus; *IPD*, inferior precentral dimple; *L*, lateral sulcus; *LU*, lunate sulcus; *OT*, occipitotemporal sulcus; *P*, principal sulcus; *PO*, parieto-occipital sulcus; *S*, spur of the arcuate sulcus; *SPD*, superior precentral dimple; *ST*, superior temporal sulcus. (Reprinted from Rizzolatti et al. 1998 with permission from Elsevier)

4.3.2 The Organizational Principle: Parieto-Frontal Circuits

As outlined above, a multiplicity of areas has also been defined in the posterior parietal cortex (Fig. 10B, C). Anatomically, the posterior parietal cortex consists of two lobules separated by the intraparietal sulcus: the superior parietal lobule (SPL) and the inferior parietal lobule (IPL). A map of the SPL and IPL is shown in Fig. 10C. The SPL and IPL receive somatosensory and visual inputs. The caudal areas of both the SPL and IPL process mainly visual information, whereas the rostral areas are related to somatosensory input in the SPL and to an integration of somatosensory and visual inputs in the IPL (Rizzolatti et al. 1998). Posterior parietal and frontal areas are connected in a specific way. Each frontal area receives afferents from a specific group of posterior parietal areas, the input from one parietal area being rich ("predominant" input) and from the other areas of the group being moderate or weak ("additional" inputs). Conversely, each posterior parietal area projects to a group of frontal areas, again the output to one frontal area being rich ("predominant" output) and to the other areas of the group being moderate or weak ("additional" outputs). The "predominant" connections between posterior parietal and frontal areas are summarized in Fig. 11. Parietal and frontal areas linked by "predominant" connections have similar functional properties. They constitute several parieto-frontal circuits working in parallel and being involved in specific sensory-motor transformations for goal-directed actions. These circuits are important functional units of the cortical motor system of primates (Rizzolatti et al. 1998).

4.3.3
The Isocortical Motor System in Macaques

The units of the macaque isocortical motor system are described in terms of their microstructure, connectivity, and functional properties in the following sequence: primary motor cortex (area F1), SMA proper and pre-SMA on the mesial cortical surface (areas F3 and F6, respectively), the PMd on the dorsolateral convexity (areas F2 and F7), and the PMv on the ventrolateral convexity (areas F4 and F5).

4.3.3.1
Primary Motor Cortex (Area F1)

Microstructure

Cytoarchitectonically, area F1 is characterized by low cell density, poor lamination, absent layer IV (inner granular layer), and very prominent giant pyramidal cells in layer V (Fig. 12A). Higher cell density, an emerging layer IV, and no giant pyramidal cells (Fig. 12B) differentiate the primary somatosensory cortex from area F1. On the lateral convexity, the border between area F1 and the somatosensory cortex lies close to the fundus of the central sulcus. The rostral border of area F1 (with areas F2, F3, and F4) is much more difficult to define since the cytoarchitectonic features do not change abruptly but rather merge gradually. The giant pyramidal cells also gradually

Fig. 11A–C Parieto-frontal circuits in the macaque monkey. Parieto-frontal projections originating from the mesial parietal cortex, free surface of the superior parietal lobule, and medial bank of the intraparietal sulcus are shown in **A** (inferior parietal lobule and occipital lobe removed), from the fundus and lateral bank of the intraparietal sulcus in **B** [intraparietal sulcus opened (*dashed line* indicates fundus) and occipital lobe removed], and from the free surface of the inferior parietal lobule in **C**. For other conventions see Fig. 10. (Reprinted from Rizzolatti et al. 1998 with permission from Elsevier)

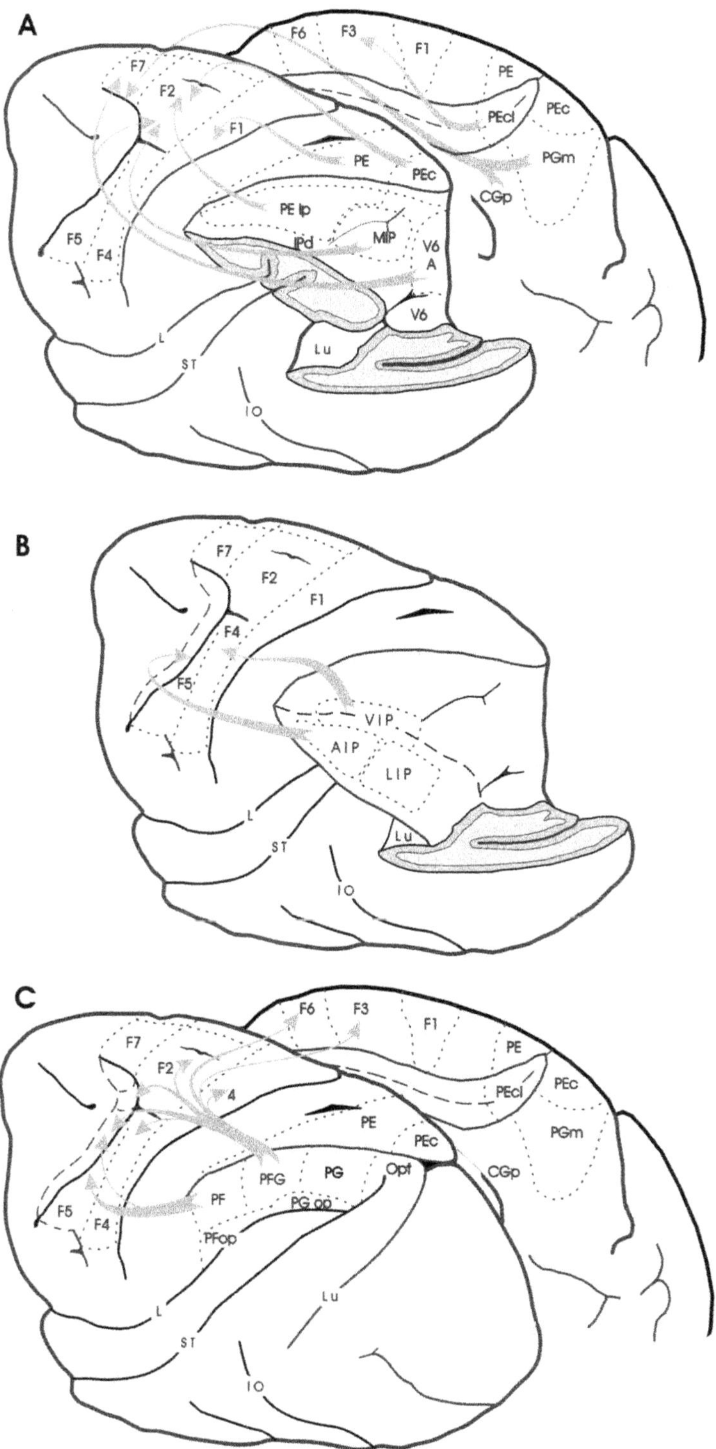
A
F6
F3
F1
PE
F7
F2
PEcl
PEc
F1
PGm
PE
PEc
CGp
PE Ip
F5
IPd
MIP
V6
F4
A
V6
L
Lu
ST
IO
B
F7
F2
F1
F4
F5
VIP
AIP
LIP
L
Lu
ST
IO
C
F6
F3
F1
F7
PE
F2
PEcl
PEc
4
PE
PGm
PEc
PFG
PG
Opt
CGp
PF
PG op
F5
F4
PFop
Lu
L
ST
IO

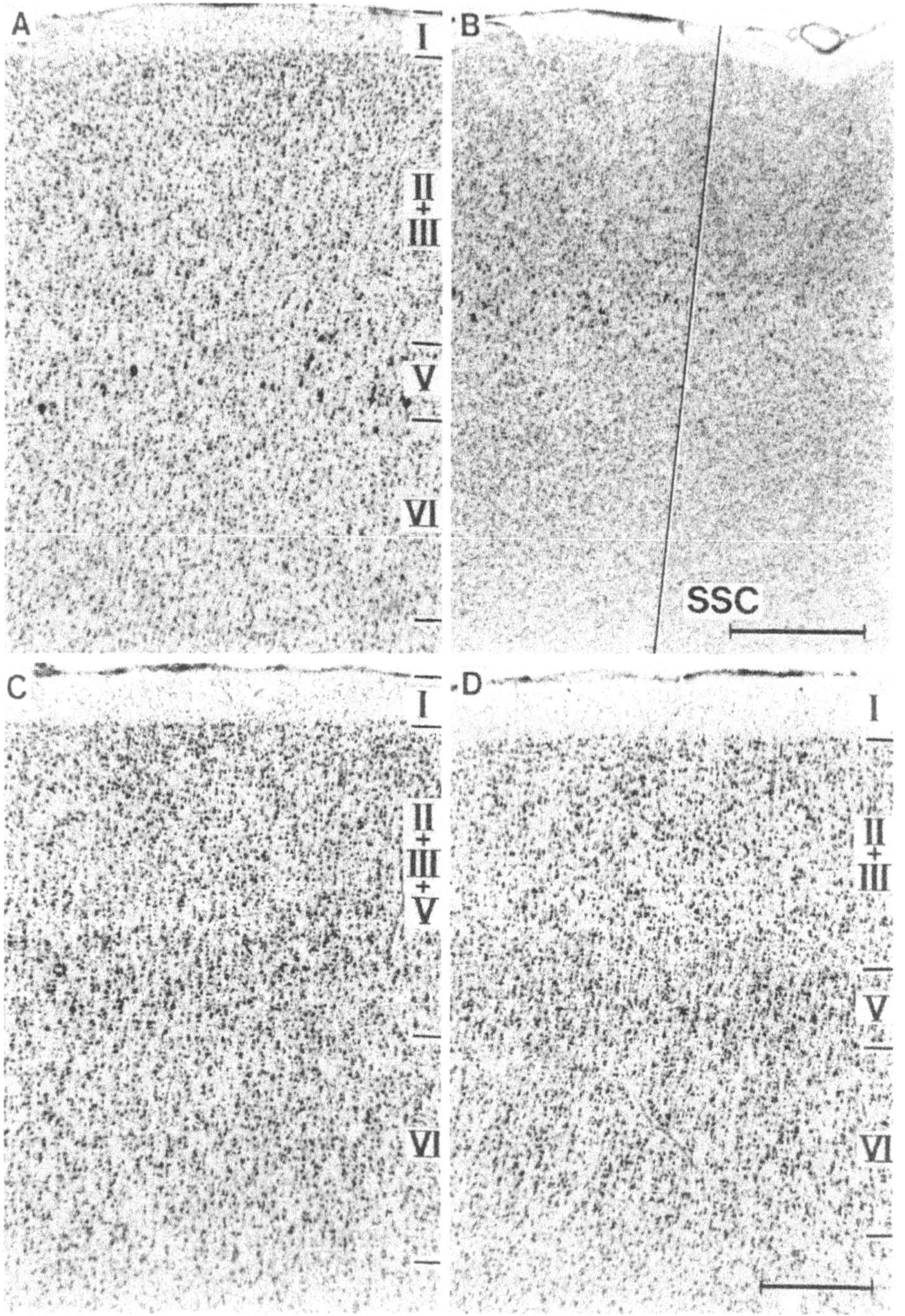

Fig. 12A–D Cytoarchitecture of area F1 (**A**), the border between area F1 and the primary somatosensory cortex (*SSC*; **B**), area F3 (**C**), and area F6 (**D**). In a caudorostral direction (from area F1 to F6), cellular density in layers *III* and *V* increases and cortical layers stand out more clearly. In the SSC, an inner granular layer (layer *IV*) emerges. *Roman numerals* indicate cortical layers. Scale bar, 500 μm (**A, C, D**) and 1 mm (**B**). (Reprinted from Geyer et al. 2000a)

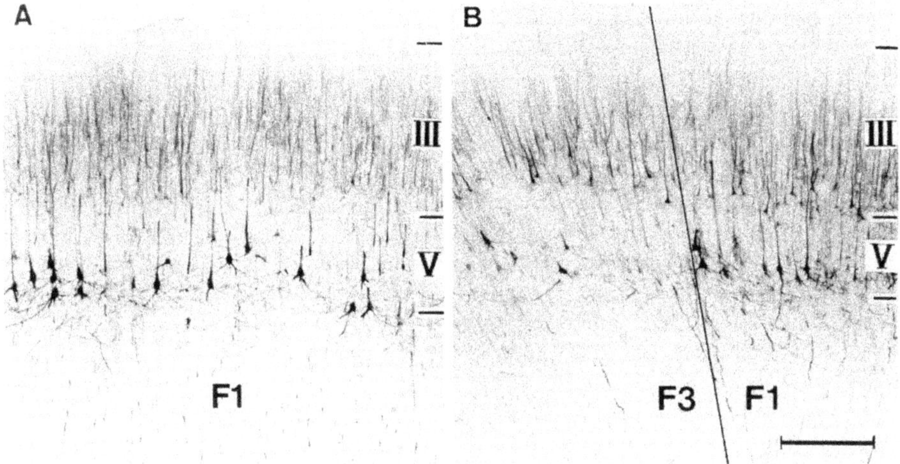

Fig. 13A, B SMI-32 immunoarchitecture of area *F1* (**A**) and the border between area *F1* and *F3* (**B**). In area *F1*, pyramidal cells in layers *III* and *V* and giant pyramids (Betz cells) in layer *V* are immunoreactive. A sharp drop in immunoreactivity marks the border between area *F1* and *F3*. *Roman numerals* indicate cortical layers. Scale bar, 500 μm. (Reprinted from Geyer et al. 2000a)

decrease in density and there is considerable interindividual variability. A more recent technique, the immunohistochemical staining of nonphosphorylated epitopes on the neurofilament protein triplet with the monoclonal antibody SMI-32 (Sternberger Monoclonals, Baltimore, USA; Lee et al. 1988; Sternberger and Sternberger 1983) labels a subset of pyramidal neurons with highly specific regional and laminar distribution patterns (Fig. 13). Neurofilament proteins are involved in the maintenance and stabilization of the cytoskeleton of the axon. Their immunohistochemical detection in the soma has been correlated with neuronal and axonal size and the conduction velocity of nerve fibers. For example, the percentage of neurofilament protein-immunoreactive pyramidal neurons is low (typically less than 50%) in short corticocortical pathways and high (typically more than 50%) in long association pathways (Hof et al. 1995). Since only a subset of pyramidal neurons is immunostained, it is easier to differentiate cortical areas in this "simplified" view of cortical architectonics. With SMI-32 immunohistochemistry, the rostral border of area F1 (Fig. 13B) can be defined more precisely, e.g., by changes in the immunoreactivity of layers III and V (Gabernet et al. 1999; Geyer et al. 1998b, 2000c). The rostral border of area F1 lies on the exposed cortical surface (Fig. 10C). Close to the midline, area F1 may reach up to the level of the caudal end of the superior precentral dimple. Further laterally, its rostral border recedes caudally and eventually disappears in the depth of the central sulcus. A small sector of area F1 occupies the mesial cortical surface (Fig. 10A). The ventral border lies in the upper bank of the cingulate sulcus.

All these data point to a microstructural homogeneity of area F1. Some years ago a subdivision of the primary motor cortex (except for the face region) into a rostral (area M1r) and a caudal (area M1c) band was proposed in the New World owl monkey (*Aotus trivirgatus*; Stepniewska et al. 1993). Differences in cytoarchitecture (layer V pyramids are smaller in M1r than in M1c), connectivity (M1r is associated with the non-primary motor and somatosensory cortex, M1c only with the somatosensory cortex), and electrophysiological properties (thresholds for eliciting movements with intracortical microstimulation are higher on average in M1r than in M1c) indicate a different involvement of the two subdivisions in the control of motor activities. It was argued that M1r may be preferentially involved in the early stages of a movement including postural adjustments to maintain balance, in reaching toward an object, and in pre-shaping the fingers to accommodate the object. M1c may be mainly involved in the later stages of a movement (when cutaneous and kinesthetic feedback is crucial) including fine adjustments of the fingers (Stepniewska et al. 1993). A similar subdivision of the primary motor cortex has been shown in humans (see below) but not in Old World macaque monkeys.

Connectivity

Although area F1 has no inner granular layer (the "classical" termination of thalamic afferents), this does not mean that area F1 receives no afferents from the thalamus. The rostral part of area F1 receives its main thalamic input from the nucleus ventralis lateralis pars oralis (VLo). Smaller contributions arise from the nucleus ventralis posterolateralis pars oralis (VPLo) and the nucleus ventralis lateralis pars caudalis (VLc). The caudal part of area F1 is innervated mainly from nucleus VPLo plus a small contribution from nucleus VLo (Matelli et al. 1989). Thalamocortical terminals are present in all layers of area F1 but they are most dense in layer III and the deep part of layer V (Sloper and Powell 1979; Strick and Sterling 1974).

Corticocortical afferents to the F1 hand area arise mainly from the SMA, followed by the lateral premotor cortex, areas 1, 2, and 5 (Ghosh et al. 1987). Callosal afferents originate from pyramidal cells in lower layer III in the homotopic part of contralateral area F1 and terminate in layers I–III (Sloper and Powell 1979). Callosal connections between the representations of the most distal parts of the arms and legs have not been described (Gould et al. 1986; Jones and Powell 1969).

Subcortical projections of area F1 originate from pyramidal cells in layer V [except for some corticostriate fibers that arise from layer III and most corticothalamic fibers that arise from layer VI (Goldman-Rakic and Selemon 1986)]. Within layer V there is a superficial-to-deep gradient of neurons that project to the striatum, midbrain, brain stem, and spinal cord (Jones and Wise 1977).

In addition to differences in their laminar origin, descending motor projections also originate from different regions of the sensorimotor cortex. In contrast to older studies (Holmes and May 1909) it is now clear that the corticospinal tract originates not only from area F1. Instead, there are substantial projections from other frontal and parietal areas. Studies based on fiber degeneration (Russel and DeMeyer 1961) and the distribution of retrogradely labeled neurons after injections of horseradish peroxidase (HRP; Murray and Coulter 1981; Toyoshima and Sakai 1982) have shown that 40–60% of the corticospinal fibers originate from the frontal lobe (30–50% from area 4; 10–30% from area 6) whereas the remaining 40–60% originate from the parietal lobe [according to the study of Toyoshima and Sakai (1982): 13% from areas 3, 1, and 2; 12% from area 5; 12% from the secondary somatosensory cortex = in toto 37% from the parietal lobe]. In a more recent study HRP conjugated with wheat germ agglutinin was injected into the lower cervical spinal cord and the distribution of corticospinal neurons was studied in the frontal lobe. Fifty percent of the labeled cells were found in the primary motor cortex, 15–20% in the cingulate motor areas, 10–20% in the SMA, and 10–20% in the lateral premotor cortex (Dum and Strick 1991).

Most interestingly, the motor capacity of a mammal (i.e., whether the distal upper extremity is used for locomotion, as a primitive hand, or for manipulation with prehensile digits and an opposable thumb) only depends on the spinal level to which the corticospinal tract descends and the presence of *direct* cortico-*motoneuronal* projections, but not on other factors, e.g., fiber number of the corticospinal tract, mean fiber diameter, or the size of the largest fibers. For further details see Geyer et al. 2000a and Porter and Lemon 1993. Direct cortico-motoneuronal projections are far more numerous in humans and the great apes than in the macaque (Kuypers 1981) and it seems that these cortico-motoneuronal connections are crucially important for independent finger movements (Bernhard et al. 1953; Kuypers 1962).

Function

Area F1 is organized somatotopically. On the cortical convexity, the face is represented laterally close to the Sylvian fissure, the leg medially close to the midline (extending on the mesial wall of the hemisphere), and the arm in between (Fig. 10A, C). Proximal parts of the extremities and axial body parts are represented mainly on the exposed cortical surface, whereas distal parts of the extremities and acral parts of the face (lips and tongue) are represented in the rostral bank of the central sulcus (Woolsey et al. 1952). Those parts of the body that are used for fine movements (fingers, lips, tongue) are represented over a larger cortical territory than are body parts that are used for coarser movements (trunk, proximal extremities). By accordingly distorting the proportions of the body, Woolsey (1952) generated the cartoon of the so-called "motor simiusculus", the monkey equivalent of the more famous "motor homunculus" by Penfield (1952). Somatotopic organization, however, should not be overinterpreted. Somatotopy means an orderly representation of the body's periphery; it does not mean a one-to-one map of individual muscles, joints, or movements. Both hand and arm muscles (Donoghue et al. 1992) and finger movements (Schieber and Hibbard 1993) are represented as overlapping and *not* as somatotopically segregated neuronal populations, thus indicating a higher-order representation that lies beyond the simple muscle or joint level (see also Poliakov and Schieber 1999). Intracortical microstimulation reveals that individual muscles are typically activated at multiple, spatially separated locations and usually more than one muscle is activated from a single site (Donoghue et al. 1992). There is *convergence of outputs* on the one hand, i.e., the outputs from a large F1 territory converge on the spinal motor neuron pool that controls muscles moving a given body part. On the other hand, there is *divergence of outputs*, i.e., the outputs from single F1 neurons ramify and synapse on multiple spinal neuron pools. The latter principle has also been confirmed anatomically. Intracellular staining has shown that a single physiologically identified corticospinal axon branches extensively within the spinal cord and terminates within the motor neuron pools of many muscles (Shinoda et al. 1981).

What does area F1 contribute to motor control? In the 1960s, Edward Evarts started to record the extracellular electrical activity from area F1 pyramidal tract neurons in awake and behaving monkeys. In one of these experiments, the monkeys were trained to raise and lower weights with flexion and extension movements of the wrist (Evarts 1968). Flexion- and extension-related neurons could be differentiated. A typical flexion-related neuron would begin to discharge as early as several hundred milliseconds before flexion onset, increase its firing rate when the beginning of the movement approached, and fire with maximal intensity during flexion. During extension, the same neuron fired with considerably lower frequency or was silent altogether. An extension-related neuron would behave in the opposite way through the flexion–extension maneuver. In subsequent studies, it could be shown that the discharge frequencies of pyramidal tract neurons were related to a number of mechanical parameters of the movements the animals were performing, e.g., the force the monkeys needed to exert (F), rate of change of force $\left(\frac{dF}{dt}\right)$, joint position, or velocity of the movement, whereby single neurons often encoded several of these parameters. Thus, it seems that area F1 neurons control kinematic and dynamic parameters of movements, whereas motor-related areas other than F1 (SMA proper, pre-SMA, areas of the PMd and PMv) seem to use external (e.g., sensory) or internal cues to trigger and guide movements (see below).

Given the fact that area F1 neurons are fairly broadly tuned in terms of movement parameters, how can we make precise movements, e.g., when reaching to a target in space? This issue was addressed by Apostolos Georgopoulos and colleagues who trained monkeys to move their hand and arm from the center of a circle toward a small light whose position varied randomly around a circle (Georgopoulos et al. 1982). The cells in area F1 fired maximally during movement in one direction (so-called preferred direction), but also (though less vigorously) during movements that deviated clockwise or counterclockwise from the preferred direction. Each neuron's directional tuning curve resembled the shape of a Gaussian curve with the curve's peak representing the neuron's preferred direction. Such a coarse tuning at the single neuron level could not explain the precise movements the monkey was performing in space. However, Georgopoulos recorded not only from one cell but from more than 200 neurons and established a tuning curve and a *direction vector* for each neuron. A neuron that fired most vigorously during a movement, say, to the left was represented by a direction vector pointing to the left. For a movement in any other direction, the same neuron's direction vector pointed to the left as well, but the length of the vector decreased proportionally to the decrease in the neuron's firing rate. For each direction of movement, all single direction vectors were plotted together and averaged to yield a so-called *population vector* for this movement direction. Most interestingly, for each movement direction, an excellent agreement was found between the direction of the actual movement the monkey was making and the direction of the population vector.

This work suggests three important conclusions about how area F1 commands voluntary movements: (1) much of the motor cortex is active for every movement, (2) the activity of each cell represents a single "vote" for a particular direction of movement, and (3) the direction of movement is determined by a tally (and averaging) of the votes registered by each cell in the population (Bear et al. 1996).

4.3.3.2
Supplementary Motor Areas "SMA Proper" and "Pre-SMA" (Areas F3 and F6)

Microstructure

The term "supplementary motor area" (SMA) was introduced by Clinton Woolsey and co-workers. Electrical stimulation of the monkey cortex revealed a complete somatotopic map of the body in the primary motor cortex ("motor simiusculus"; see above) and an additional ("supplementary") complete motor representation further rostrally on the mesial cortical surface (Woolsey et al. 1952). For a long time the SMA was considered a homogeneous entity. However, over the last years it has become clear that the SMA in fact corresponds to two areas that differ from each other both in terms of microstructure and function, namely, SMA proper (or area F3) and pre-SMA (or area F6).

From a historical perspective, it is not a new concept that the non-primary motor cortex on the mesial surface is not homogeneous. The Vogts (1919) subdivided this region into areas 6aα and 6aβ, von Bonin and Bailey (1947) into areas FB and FC, and Barbas and Pandya (1987) into areas 6DC and MII (cf. Table 3). However, these maps vary considerably in terms of size and rostrocaudal extent of the areas. In addition, most of these areas extend from the mesial surface further laterally on the dorsolateral convexity up to the level of the superior branch and spur of the arcuate sulcus. The functional properties of neurons on the dorsolateral convexity, however, are markedly different from those on the mesial cortical surface—a fact that is not taken into account by these maps.

The parcellation of the mesial cortex into areas F3 and F6 is based on histochemical (Matelli et al. 1985), cytoarchitectonic (Matelli et al. 1991), and, most recently, receptor autoradiographic data (Fig. 14; Geyer et al. 1998a). Clear-cut changes in the staining pattern of cytochrome oxidase and in the laminar binding patterns especially of glutamate (revealed with [^{3}H]α-amino-3-hydroxy-5-methyl-4-isoxalone propionic acid (AMPA) and [^{3}H]kainate; Fig. 14C–F) and muscarinic cholinergic (revealed with [^{3}H]oxotremorine-M) binding sites very closely match corresponding cytoarchitectonic borders. In Nissl-stained sections, area F3 is poorly laminated like area F1 (Fig. 12C). The feature most distinguishing area F3 from area F1 is an increase in cellular density in lower layer III and upper layer V. Scattered giant pyramidal cells are present only in the caudal part of area F3 toward the border with F1. Unlike areas F1 and F3, area F6 is clearly laminated (Fig. 12D). Its most prominent feature is a dark layer V well demarcated from layers III and VI. At the rostral border of area F6 (with the prefrontal cortex), an incipient layer IV becomes evident. Area F3 lies on the mesial cortical surface immediately rostral to area F1 and extends for 8–10 mm in the rostrocaudal direction (Fig. 10A). Area F6 lies anterior to area F3. It has a common border with area F3 and extends rostrally for 5–6 mm up to the granular prefrontal cortex. No macroanatomical landmarks indicate the borders between areas F1 and F3, F3 and F6, and F6 and the prefrontal cortex. Areas F3 and F6 abut the cingulate cortex approximately in the middle of the dorsal bank of the cingulate sulcus and extend on the dorsolateral convexity for approximately 2–3 mm (Fig. 10C; Geyer et al. 1998a; Matelli et al. 1991).

Connectivity

The different functional roles of areas F3 and F6 in motor control are further substantiated by their different input and output properties. Area F3 receives its main thalamic input from the nucleus ventralis lateralis pars oralis (VLo), whereas the input to area F6 comes mostly from the nucleus ventralis anterior pars parvocellularis (VApc), area X of Olszewski (1952), and the nucleus medialis dorsalis (Matelli and Luppino 1996; Rizzolatti et al. 1996c). Both areas are elements of different subcortical motor loops. Area F3 is the target (via nucleus VLo) of the putamen and pallidum,

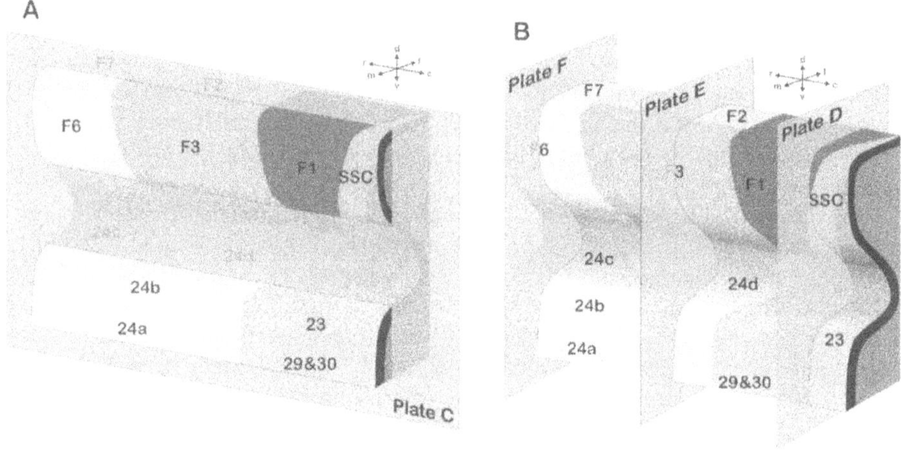

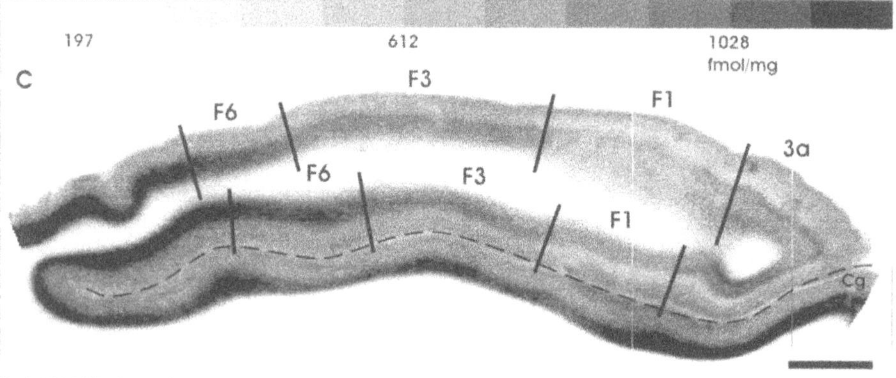

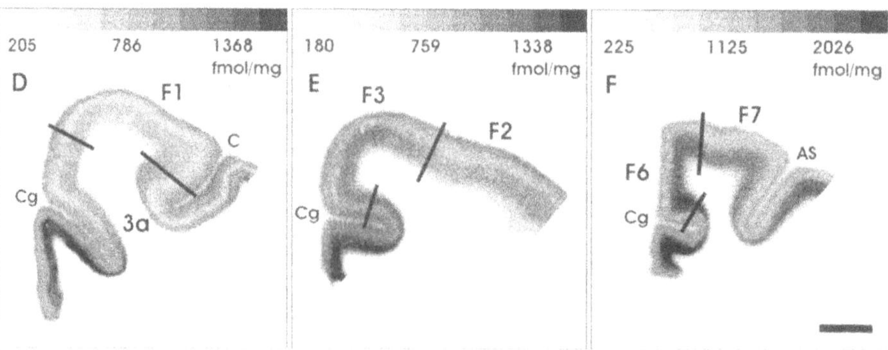

Fig. 14 Schematic drawing of the mesial cortical surface with the plane of sectioning in *plate* **C** (**A**) and *plates* **D**, **E**, and **F** (**B**). The cingulate sulcus is opened to expose the areas in its dorsal and ventral bank. Numerals *23, 24a, 24b, 24c, 24d, 29, and 30* are areas of the cingulate cortex (see also Matelli et al. 1991) and *SSC* is the primary somatosensory cortex. Directionally: *r*, rostral; *c*, caudal; *m*, medial; *l*, lateral; *d*, dorsal; *v*, ventral. **C–F** Contrast-enhanced autoradiographs of [^{3}H]kainate binding sites in a paramedian section (**C**) and a series of coronal sections (**D–F**). *Light gray* indicates low densities, *dark gray* high densities of binding sites (cf. bar). B_{max} values are expressed as "fmol/

whereas area F6 receives its input (via nucleus VApc) from the caudate (Alexander et al. 1986) and the cerebellum (Rouiller et al. 1994).

Cortical afferents to area F3 originate predominantly from areas F2 and F4 (~25%), areas F5, F6, and F7 (~20%), primary and secondary somatosensory cortex, posterior parietal areas PE and PEci (Fig. 10A, C; ~20%), cingulate cortex (~20%), and primary motor area F1 (~15%). Area F6 is mainly connected with areas F5 and F7 (~40%), prefrontal cortex (~20%), cingulate cortex (~20%), areas F2, F3, and F4 (~15%), posterior parietal areas PG, PFG (Fig. 10C), and the superior temporal sulcus (~5%). In contrast to area F3, F6 is *not* connected with area F1 (Luppino et al. 1993; Rizzolatti et al. 1996c).

Descending projections from area F3 are corticospinal and corticobulbar in nature. The caudal part of area F3 projects to the thoracolumbar segments and the rostral part to the cervicothoracic segments of the spinal cord. Only corticobulbar projections originate from area F6 (He et al. 1995; Keizer and Kuypers 1989; Luppino et al. 1994; Rizzolatti et al. 1996c).

Function

Area F3 is somatotopically organized. The arm and leg representations are two oblique bands running in a dorsorostral-to-ventrocaudal direction. The arm is represented in the rostral, the leg in the caudal band (Fig. 10A). A small orofacial representation is located even further rostrally at the border with area F6. Area F6 contains only a representation of the arm (Luppino et al. 1991).

Intracortical microstimulation (ICMS) experiments revealed marked differences between the movements evoked from mesial area F1, area F3, and area F6. Almost all stimulation sites (99%) in mesial area F1 are excitable with standard ICMS procedures. Movements can be evoked at low thresholds (on average <20 μA) and are typically fast and short-lasting in nature. Most movements (~90%) are simple (i.e., restricted to a single joint or the digits of one extremity), whereas only ~10% of the induced movements are contiguous (i.e., displacements occurring at two adjacent joints) or complex (i.e., displacements of more than two articulations or of noncontiguous articulations or body parts, e.g., shoulder and wrist). Proximal and distal movements are equally well represented and are spatially clearly segregated. In contrast, in area F3 only ~80% of the stimulation sites are excitable with standard ICMS procedures. The evoked movements are typically fast and short-lasting as well, but movement thresholds are higher (on average 10–30 μA) than in area F1. The nature of the induced movements is also somewhat different from mesial area F1. In area F3 only ~60% are simple movements (vs. ~90% in mesial F1), whereas ~30% are complex (vs. ~5% in mesial F1). Proximal movements are much better represented in area F3 than are distal movements and they are spatially poorly separated. Finally, in area F6, only ~20% of the stimulation sites are excitable with standard ICMS proce-

mg protein". Changes in the distribution pattern of binding sites that coincide with cytoarchitectonically defined borders in adjacent cell-stained sections are marked. The primary somatosensory area 3a is designated *3a*; *AS*, superior arcuate sulcus; *C*, central sulcus; *Cg*, cingulate sulcus. Scale bar, 5 mm (**C–F**). (Reprinted from Geyer et al. 2000a) dures and the movement thresholds are even higher (on average 20–40 μA) than in area F3. Sixty percent of the induced movements are fast and short-lasting, whereas 40% are slow displacements of the limbs that somewhat mimic natural movements or postural adjustments of the animal. This latter type of movements cannot be evoked from area F3 or mesial area F1. In area F6, proximal movements are also much better represented than are distal movements and spatial separation is also poor (Luppino et al. 1991). To sum up, the ICMS data show that from mesial area F1 to area F6 cortical excitability decreases whereas the thresholds at which movements can be elicited and the degree of complexity of the induced movements increase.

What does the SMA (SMA proper and pre-SMA) contribute to motor control? Traditionally it was thought that the SMA is more involved in the control of complex (versus simple) motor tasks, in the internal (versus external) initiation of motor acts, and in controlling proximal (versus distal) limb movements. A recent review of "new concepts of the supplementary motor area" (Tanji 1996) however concluded that the SMA influences many aspects of motor behavior and over-simplification of the "function" of the SMA is no longer appropriate. The main concepts are briefly summarized here (cf. Tanji 1994, 1996).

Complex Versus Simple Motor Tasks. What does "simple" or "complex" mean? "Simple" does not necessarily mean "a small body part is involved in the movement." It is by no means "simple" to move one finger without moving others since this involves active suppression of many finger muscles. Nor is a movement, even if it is apparently easy to execute, "simple" if it must be selected according to complex rules. "Simple" in this context means performing a "kind of motor task that does not include a temporal or spatial structure imposing a great deal of specific requirements for the subjects" (Tanji 1994). Neuronal activity in the SMA changes before and during execution of "simple" movements (Brinkman and Porter 1979; Okano and Tanji 1987; Romo and Schultz 1987; Tanji and Kurata 1979, 1982; Thaler et al. 1988). On the other hand, the effects of unilateral or bilateral lesions of the SMA appear surprisingly mild at first glance. No gross motor deficits were observed. The animals were able to run, climb, and grasp food without major difficulties (Brinkman 1984; Passingham et al. 1989). However, subtle effects became evident when the monkeys performed more complex hand movements. Brinkman (1984) observed a transient clumsiness when food was retrieved unimanually with relatively independent finger movements and a long-lasting failure in bimanual coordination (i.e., the two hands tended to behave in a similar manner instead of sharing the task between them). Halsband (1982) noted that after an SMA lesion the monkeys could not perform a sequence of three movements with a manipulandum. Thus, the old concept is no longer appropriate since the SMA is involved in both "simple" and "complex" motor tasks.

Internal Versus External Motor Initiation. What is "internal" and what is "external" initiation of movements? Passingham (1993) gives a simple yet very graphic example. "Consider a dog lying on the floor. It comes when its master calls; but there are other occasions on which it gets up and comes of its own accord. In the first case there is a change in the environment: it hears the call. In the second case there is no change in the environment: instead there is a change in the animal itself. In one instance the animal reacts to an external event, and in the other its action is self-initiated. We may say that the individual 'acts'." A traditional concept has been that the SMA is

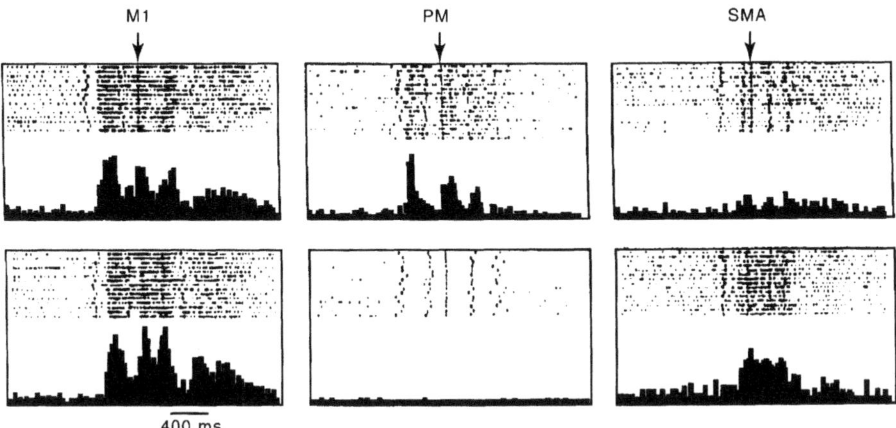

Fig. 15 Extracellularly recorded activity from a neuron in the primary motor cortex (*M1; left column*), lateral premotor cortex (*PM; middle column*), and supplementary motor area (*SMA; right column*) when a monkey pressed three buttons in sequence. In the first trial (*top row*), the sequence was externally (i.e., visually) cued by lighting the buttons. In the second trial (*bottom row*), the same sequence was internally cued. The *M1* neuron showed similar activity whether the animal performed from external or internal cues. The *PM* neuron was much more active in response to external than internal cues, while the opposite was true for the *SMA* neuron. (Reprinted from Schieber 1999 with permission from Elsevier)

involved in self-initiated or internally guided movements, whereas the lateral premotor cortex is involved in movements triggered or guided by sensory signals (Eccles 1982; Goldberg 1985; Rizzolatti et al. 1983; Tanji 1994). Subsequent electrophysiological studies, however, have shown that neurons both in the SMA and the lateral premotor cortex are active no matter whether monkeys performed self-initiated ("internal") or externally guided ("external") movements (Kurata and Wise 1988; Okano and Tanji 1987; Romo and Schultz 1987; Thaler et al. 1988). There is no simple dichotomy in function between the SMA and the lateral premotor cortex (at most there is a *relative* importance of the SMA in self-initiated and the lateral premotor cortex in externally guided movements; cf. Fig. 15). Recent studies have confirmed this concept: the SMA is involved both in externally guided, e.g., visually triggered motor tasks (Grafton et al. 1996b) and in self-initiated motor acts, e.g., in the temporal sequencing of multiple movements without sensory guidance (Tanji and Shima 1994) or in the poor ability of SMA-lesioned monkeys to select appropriate movements when no sensory cue is available (Chen et al. 1995; Thaler et al. 1995).

Proximal Versus Distal Limb Movements. The old notion that the SMA is principally involved in proximal movements originated mainly from ICMS studies (see above) that found proximal movements to be much more fully represented than distal movements. Unit recording studies, however, have detected neuronal activity in the SMA associated both with proximal and distal movements (Brinkman and Porter 1979; Tanji and Kurata 1979). Even when the movements were specifically limited to the hand, neuronal activity has been observed in the SMA (Okano and Tanji 1987; Tanji et al. 1988; Thaler et al. 1988). How can this conflict between ICMS and unit recording data be resolved? Tract tracing studies have yielded conflicting results. Luppino et al. (1993) injected tracers into the arm field of the SMA and found retrogradely labeled neurons in area F1 almost exclusively on the free cortical surface (representation of the proximal arm; see above), whereas Tokuno and Tanji (1993) found that the F1 representations of the proximal arm (on the cortical convexity) *and* the fingers (in the rostral bank of the central sulcus) are connected with the SMA. In a recent review article (1994), Tanji pointed out that ICMS may yield misleading results in non-primary motor areas. When motor effects are to be evoked from secondary motor areas, higher electrical currents or longer trains of stimuli are needed. When using such parameters, motor effects are often elicited in multiple muscles, giving rise to complex multijoint movements and masking motor effects in smaller distal muscles. In addition, the stimuli induced by ICMS are rather artificial in nature and this should caution against an overinterpretation of the outcome of high-frequency multiple-pulse electrical stimulation with respect to the physiological organization of a non-primary motor area (Tanji 1994). Hence, the SMA seems to be involved in proximal and distal movements and the old concept should be discarded.

Motor Preparation. Tanji et al. (1980) instructed monkeys to push or pull a handle depending on a forthcoming trigger signal. During the preparatory period, the neuronal activity in the SMA increased or decreased depending on whether the monkey was intending to push or pull the handle. The preparation for different movements was reflected by differential neuronal activity. These findings could be replicated in subsequent studies (Alexander and Crutcher 1990; Kurata and Wise 1988). However, preparatory activity for movements in different directions is found in many other cortical areas, including area F1, lateral premotor cortex, prefrontal cortex, parietal cortex, and basal ganglia (see Tanji 1994 for references). One may wonder, whether there are any paradigms that describe more specifically the role that the SMA plays in the preparation for forthcoming movements. This seems to be the case for motor preparation that implies a greater degree of complexity. Differences in activity was found in the SMA when an additional instruction signal was introduced that instructed the monkey what to do when the trigger signal came up (i.e., instruction signal A → press a key in response to trigger signal X, but withhold the movement in response to Y; instruction signal B → do just the opposite; Tanji and Kurata 1985), when the animal prepared to move a single hand or both hands (Tanji et al. 1988), or when the monkey prepared to make sequential movements under sensory guidance or based on memory (Mushiake et al. 1991).

Bimanual Coordination. Brinkman (1981, 1984) evaluated the influence of a unilateral SMA lesion on bimanual hand coordination in monkeys. Monkeys had to retrieve a food pellet lodged in a hole in a perspex sheet with both hands. A normal animal pushed from above with the index finger of the preferred hand, while the non-preferred hand was cupped under the hole, anticipating the catch of the falling pellet. A monkey with a unilateral SMA lesion (opposite the non-preferred hand) used both hands in a mirror-like way, pushing the food from above with one hand and from below with the other hand. The findings were interpreted that in a lesioned animal the intact SMA would influence the motor outflow from both sides, whereas in a normal animal the SMA on one side would inform the opposite side of intended or ongoing movements in one side. Tanji et al. (1987, 1988) trained monkeys to press a key with the right, left or both hands simultaneously depending on which instruction signal was given prior to a trigger signal. Neuronal activity in the SMA changed in relation to the movement of only one hand (right or left) or both hands together. Taken together, the findings suggest that the SMA is involved in a process of transformation of information about the intended usage of both hands into the pattern of neural activity for movement execution (Tanji 1994).

Differences in Function Between Areas F3 (SMA Proper) and F6 (Pre-SMA). In recent years, the subdivision of the "classical" SMA into a caudal (SMA proper or area F3) and a rostral (pre-SMA or area F6) region has gained wide acceptance in the neuroscience community. For this reason, there is now an increasing number of studies available that try to elucidate functional differences between the two areas. Single unit recording studies have shown that neurons responding to visual stimuli are found mainly in the pre-SMA (Matsuzaka et al. 1992), whereas neurons in the SMA proper respond mainly to somatosensory stimuli (Hummelsheim et al. 1988; Matsuzaka et al. 1992). When performing a motor task, set-related neuronal activity long before the onset of a movement is present almost exclusively in the pre-SMA (Alexander and Crutcher 1990; Matsuzaka et al. 1992; Rizzolatti et al. 1990), whereas phasic activity, more frequently time-locked to the movement onset is more common in the SMA proper (Alexander and Crutcher 1990; Matsuzaka et al. 1992). In addition, pre-SMA neurons show more complex firing patterns related to arm movements. Their activity is influenced by the distance of the objects from the monkey and frequently the neuronal activity depends on whether the animal can or cannot reach and grasp the object that is presented (Rizzolatti et al. 1990). In two recent studies monkeys were trained to perform three different movements sequentially in a temporal order. Neuronal activity was found mainly in the pre-SMA (but only infrequently in the SMA proper and not in the primary motor cortex) when the animals were required to discard a current motor plan and develop a new plan appropriate for the next movement (Matsuzaka and Tanji 1996; Shima et al. 1996). These findings are in line with the massive input that the pre-SMA (but not the SMA proper) receives from the prefrontal cortex (see above). This prompted Rizzolatti and co-workers (1998) to argue that the pre-SMA represents a "supramotor" control center that triggers a movement only when external contingencies (close or distant stimulus, presence or absence of an obstacle, physical possibility to act) and motivational factors allow it.

In summary, the SMA is involved in many different aspects of planning and executing voluntary movements and it would be too simplistic to attribute its "function" to one or a few "categories" of cortical motor control. Studies in nonhuman primates suggest that the pre-SMA (area F6) seems to be more involved than the SMA proper (area F3) in planning or acquiring a spatiotemporal pattern of movements when a subject needs to accomplish a motor task incorporating a novel requirement imposed by changes in the environmental conditions (Tanji 1996).

4.3.3.3
Dorsolateral Premotor Cortex (PMd; Areas F2 and F7)

Microstructure

Similar to areas F3 and F6 on the mesial cortical surface, different investigators have proposed various maps of the PMd that are also plagued by the subjective nature of microstructural parcellation. Most authors have subdivided the PMd into two areas, a caudal one and a rostral one. The position of the border between the two areas, however, is highly variable. It was placed at a level just rostral to the genu of the arcuate sulcus by the Vogts (1919), further rostrally (toward the prefrontal cortex) by von Bonin and Bailey (1947), or further caudally (toward the primary motor cortex) by Barbas and Pandya (1987).

In contrast to these maps, the parcellation of the PMd into a caudal area F2 and a rostral area F7 is supported by converging results from histochemical (Matelli et al. 1985), cytoarchitectonic (Matelli et al. 1991), receptor autoradiographic (Geyer et al. 1998a), and most recently immunohistochemical data on the distribution of neurofilament proteins with monoclonal antibody SMI-32 (Fig. 17; Gabernet et al. 1999; Geyer et al. 1998b, 2000c; Petrides et al. 2000). In a Nissl-stained section, area F2 is poorly laminated like areas F1 and F3 (Fig. 16A). Similar to area F3, scattered giant pyramidal cells are present only in its caudal part toward the border to area F1. A narrow band of medium-sized pyramids can be found at the border between layers III and V. Cell density in layers III and V is slightly lower than in area F3. Area F7 is clearly laminated and layer V is prominent (as in area F6; Fig. 16B). Cell density in area F7 is comparable to area F6 but the pyramidal cells are somewhat smaller. Despite these obvious differences in cytoarchitecture, the borders between the two areas are very difficult to define in Nissl-stained material, since at a border the cytoarchitectonic features tend to merge gradually and not to change abruptly in a step-like fashion. The advent of a new technique, immunostaining of neurofilaments with antibody SMI-32, has greatly facilitated a more precise and reliable definition of interareal borders. With SMI-32 immunohistochemistry, differences in size, shape, and packing density of immunopositive layer III and V pyramidal cells define the borders of areas F2 and F7 more clearly than do differences in cytoarchitecture (Geyer et al. 1998b, 2000c). Arcuate spur and superior arcuate sulcus separate areas F2 and F7 from the PMv and the prefrontal cortex (Fig. 10C). Area F2 extends 2–3 mm laterally from the midline to the fundus of the arcuate spur and the fundus of the caudal part of the superior arcuate sulcus. The border between areas F2 and F7 lies slightly anterior to the genu of the arcuate sulcus. Area F7 occupies the rostral part of the PMd 2–3 mm laterally from the midline to the fundus of the rostral part of the superior arcuate sulcus. Recent evidence from SMI-32 immunohistochemistry indicates that the fine-grained functional organization within area F2 (see below) may be reflected on a structural level: differences in layer V immunoreactive neurons define a dorsal (area F2d) and a ventral (area F2v) region (Fig. 17A). The border between areas F2d and F2v lies at the level of the superior precentral dimple and cannot be detected cytoarchitectonically in Nissl-stained material (Geyer et al. 1998b, 2000c).

Connectivity

The thalamic input to area F2 originates from the nucleus ventralis posterolateralis pars oralis (VPLo), nucleus ventralis lateralis pars caudalis (VLc), and nucleus ven-

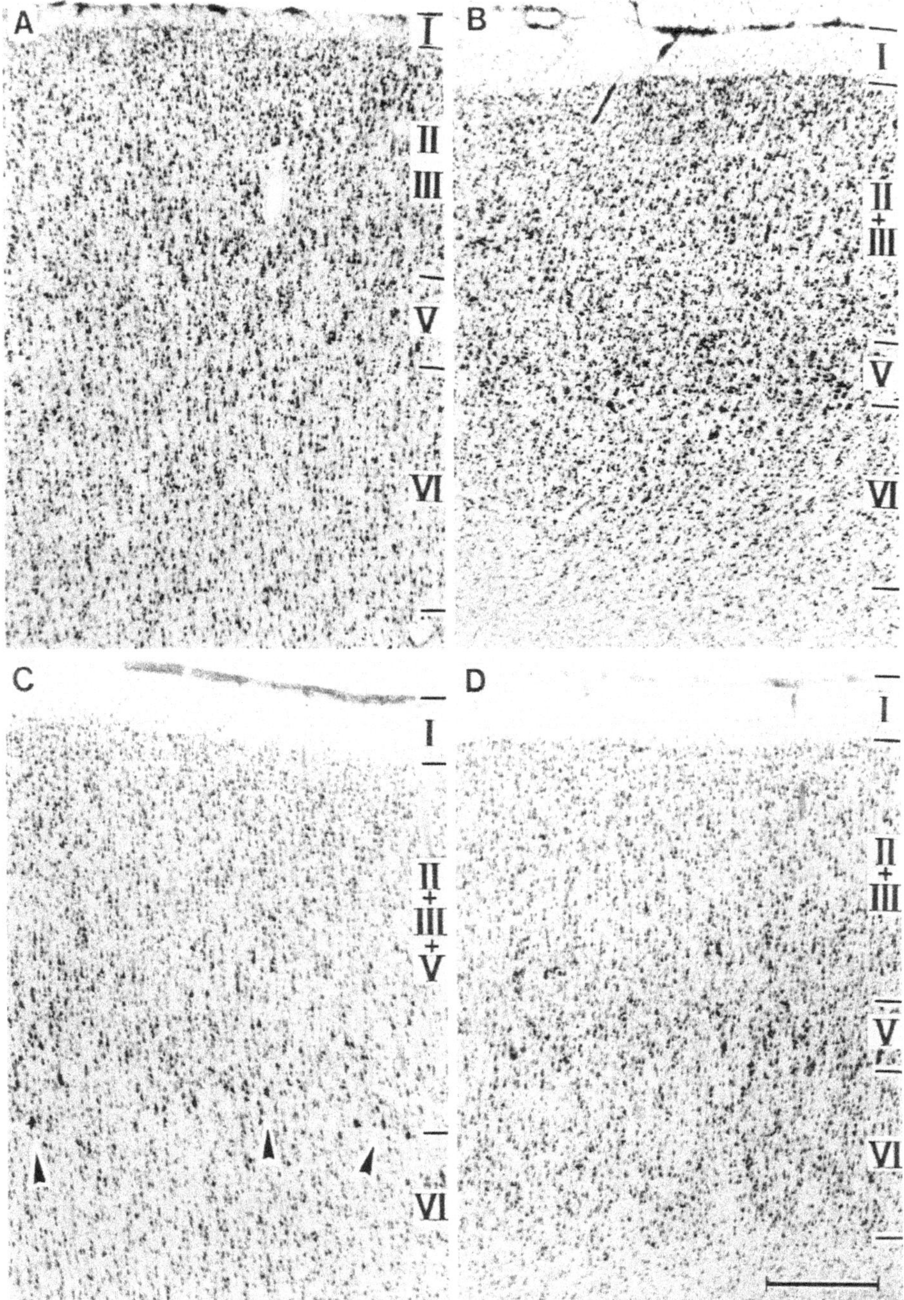

Fig. 16A–D Cytoarchitecture of area F2 (**A**), F7 (**B**), F4 (**C**), and F5 (**D**). As on the mesial surface, cellular density in layers *III* and *V* increases and cortical layers stand out more clearly in a caudorostral direction (from area F2 to F7 and F4 to F5). *Arrowheads* in **C** mark large layer *V* pyramidal cells scattered throughout the caudal and dorsomedial part of area F4. *Roman numerals* indicate cortical layers. Scale bar, 500 μm. (Reprinted from Geyer et al. 2000a)

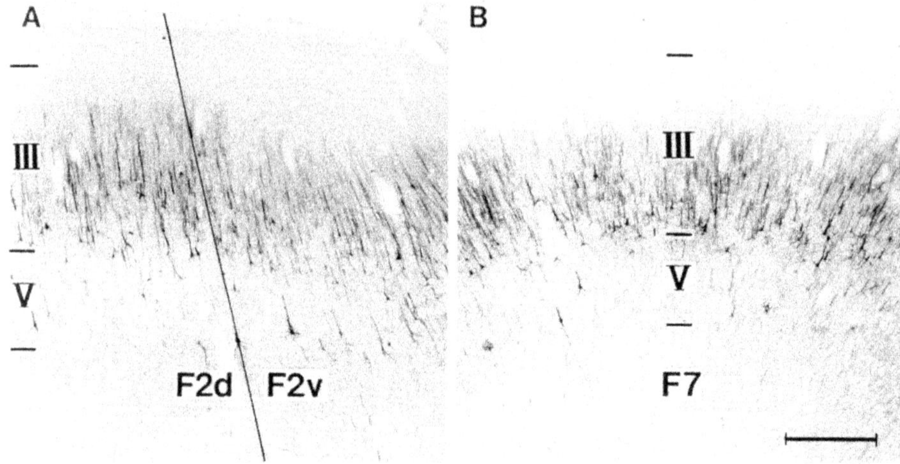

Fig. 17A, B SMI-32 immunoarchitecture of the border between area F2d and F2v (**A**) and area F7 (**B**). Pyramidal cells in layers *III* and *V* are immunopositive. Emerging small immunoreactive pyramids in layer *V* mark the border between area F2d and F2v. In area F7, cell bodies of layer *III* pyramids are slightly larger. *Roman numerals* indicate cortical layers. Scale bar, 500 μm. (Reprinted from Geyer et al. 2000a)

tralis lateralis pars oralis (VLo). The dorsal part of area F7 [supplementary eye field (SEF), see below] receives its thalamic input mainly from area X of Olszewski (1952) and nucleus ventralis anterior pars parvocellularis (VApc) and magnocellularis (VAmc). Ventral area F7 is connected with VApc, area X, VLc, and VPLo (Matelli and Luppino 1996). Similar to areas F3 and F6, area F2 and dorsal and ventral area F7 are elements of different subcortical motor loops. Areas F3 and F2 seem to belong to the same loop, being the target (via nucleus VLo) of the putamen and pallidum, whereas area F6 and ventral area F7 belong to another loop, receiving their input (via nucleus VApc) from the caudate nucleus. Dorsal area F7 (SEF) is part of a subcortical "oculomotor circuit" (Matelli and Luppino 1996).

Area F2 is connected with area F1, but not with the prefrontal cortex. Conversely, area F7 is strongly connected with the prefrontal cortex but there are no direct connections with area F1 (Barbas and Pandya 1987; Luppino et al. 1990). Areas F2 and F7 receive rich cortical input from the posterior parietal cortex. These connections are the anatomical basis of several parieto-frontal circuits that are (as described above) important functional units of the cortical motor system (Matelli et al. 1998; Rizzolatti et al. 1997, 1998). The region of area F2 around the superior precentral dimple receives its "predominant" parietal input from area PEip in the medial bank of the intraparietal sulcus and from area PEc in the caudal part of the superior parietal lobule (*"PEip/PEc–F2 dimple" circuit*; Figs. 10B, 10C, 11A). The medial intraparietal (MIP) area in the medial bank of the intraparietal sulcus and area V6A in the parieto-occipital sulcus are connected with the ventrorostral part of area F2 (*"MIP/V6A–F2 ventrorostral" circuit*; Figs. 10B, 11A). Area PGm on the mesial surface of the superior parietal lobule projects predominantly to ventral area F7 (*"PGm–F7 ventral" circuit*; Figs. 10A, 11A). Dorsal area F7 (SEF) is the target of the lateral intraparietal (LIP) area in the lateral bank of the intraparietal sulcus (*"LIP–F7 dorsal" circuit*; Fig. 10B; Huerta and Kaas 1990).

Area F2 sends descending projections to the spinal cord. Area F7 projects to the superior colliculus [mainly from dorsal area F7 (SEF)] and to the reticular formation in the brain stem (Fries 1985; He et al. 1993; Keizer and Kuypers 1989).

Function

Since the circuits connecting the parietal cortex with areas F2 and F7 are not only anatomical but also functional entities, the discussion of the functional aspects of areas F2 and F7 will be based on a more detailed description of the functional properties of these parieto-frontal loops.

"PEip/PEc–F2 Dimple" and "MIP/V6A–F2 Ventrorostral" Circuit. Neurons recorded from the part of area PE lying in the medial bank of the intraparietal sulcus [area PEip = the rostral part of area PEa of Pandya and Seltzer (1982) = that part of area PEa that gives rise to corticospinal projections (Matelli et al. 1998)] respond to somatosensory stimuli (Iwamura and Tanaka 1996; Mountcastle et al. 1975) often in association with arm movements (Kalaska et al. 1990). Neurons in the caudal part of the superior parietal lobule (area PEc) and the rostrally adjoining area PE (with which area PEc is heavily connected; Pandya and Seltzer 1982) are also involved in the processing of somatosensory stimuli for movement organization (Colby and Duhamel 1991; Galletti et al. 1996). It seems that areas PEip and PEc are mainly involved in somatosensory aspects of cortical motor control.

The situation is different in area MIP (Colby et al. 1988) in the medial bank of the intraparietal sulcus (caudal to area PEip) and in area V6A (Galletti et al. 1996) in the parieto-occipital sulcus. Neurons in area MIP respond to somatosensory and visual stimuli (Colby and Duhamel 1991), whereas neurons in area V6A respond to visual stimuli but also in association with eye or arm movements (Galletti et al. 1996, 1997). In contrast to areas PEip and PEc, areas MIP and V6A seem to be involved in somatosensory and visual aspects of motor control.

Areas PEip and PEc (mainly somatosensory information) project to the region of area F2 around the superior precentral dimple, areas MIP and V6A (somatosensory and visual information) project to the ventrorostral part of area F2.

Area F2 is electrically excitable and has a rough somatotopic organization. Leg movements are represented dorsal to the superior precentral dimple, and arm movements ventral thereto (Fig. 10C). Proximal and distal movements are poorly separated (Dum and Strick 1991; Godschalk et al. 1995; He et al. 1993; Kurata 1989). Unit recording studies confirm the differential projections of the two parieto-F2 circuits. Responses to sensory stimuli can be found throughout area F2. However, visually driven neurons are mainly concentrated in the ventrorostral sector of area F2 (Fogassi et al. 1999). By introducing a delay period between the instruction signal and the movement trigger signal, area F2 neurons can be grouped into three different classes: (1) neurons exhibiting a visually driven phasic response immediately after the instruction signal (signal-related neurons), (2) neurons showing sustained activity during the delay period between the instruction and the trigger signal (set-related neurons), and (3) neurons firing between the trigger signal and the movement onset (movement-related neurons). These three classes of neurons are distributed along a gradient within area F2, i.e., signal-related neurons are predominant in the rostral part of area F2 (toward area F7), set-related neurons in the central part, and movement-related neurons in the caudal part (toward area F1; Caminiti et al. 1996; Johnson et al. 1996; Tanne et al. 1995).

In conclusion, the "PEip/PEc–F2 dimple" circuit appears to be involved in planning and controlling arm (and leg) movements on the basis of somatosensory information. In contrast, the "MIP/V6A–F2 ventrorostral" circuit uses somatosensory and visual information, probably for the same purpose. Monitoring and controlling arm position during the transport phase of the hand toward the target could be one of the major functions of this circuit (Rizzolatti et al. 1998).

"PGm–F7 Ventral" Circuit. Neurons in area PGm (Pandya and Seltzer 1982) on the mesial surface of the superior parietal lobule fire during arm and/or eye movements (Ferraina et al. 1997a,b).

Area PGm projects predominantly to the ventral part of area F7. Neurons in area F7 respond to arm movements or to visual stimuli. In contrast to area F2, visually responsive neurons in area F7 are more numerous and they do not need a subsequent movement in order to become active (di Pellegrino and Wise 1991). When tested in a paradigm in which reaching movements were (1) triggered by and directed toward the same visual (or acoustical) target or (2) triggered by a sensory cue but directed toward a different spatial target, some area F7 neurons responded only in the first condition (Vaadia et al. 1986). Lesion studies suggest a crucial involvement of area F7 (and possibly rostral area F2 as well) in stimulus–response associations. Monkeys were trained to perform a conditional association task, e.g., to make arbitrary goal-directed movements in response to colored stimuli. After the lesion, no obvious motor deficit was present but the animals were no longer able to perform the task (Halsband and Passingham 1982; Petrides 1982).

In conclusion, the "PGm–F7 ventral" circuit appears to be important for conditional movement selection and for the visual localization of stimuli in space as a prerequisite for reaching movements.

"LIP–F7 Dorsal" Circuit. Neurons in area LIP (Andersen et al. 1985; Blatt et al. 1990) in the lateral bank of the intraparietal sulcus encode visual signals (Robinson et al. 1978) that are related to memory (Gnadt and Andersen 1988) and to saccades (rapid eye movements) toward the target (Barash et al. 1991; Goldberg et al. 1990). Neuronal activity can be modulated by attention (Bushnell et al. 1981) and by the position of the eye in the orbit (Andersen et al. 1990). More recent studies have shown that area LIP contributes to a stable internal representation of space by dynamic updating or remapping visual information in combination with eye movements (Colby et al. 1995; Duhamel et al. 1992). This may be a neural basis of the everyday experience that despite constant movement of our eyes (and thus activation of different retinal neurons) we perceive a stable visual world.

Area LIP projects to the dorsal part of area F7 that contains the SEF. In contrast to area F2, F7 is scarcely excitable electrically (Godschalk et al. 1995). An exception is the SEF from which converging or goal-directed saccades can be evoked (Schlag and Schlag-Rey 1987). Neurons in the SEF [and in the frontal eye field (FEF)] fire before a saccade is initiated (Hanes et al. 1995). However, neuronal activity in the SEF is higher than in the FEF when monkeys learn new associations between visual cues and the direction of saccades (Chen and Wise 1995). The SEF seems to be more involved than the FEF in learning-related processes. An interesting new concept has been introduced by Olson and Gettner (1995, 1996). Monkeys were trained to perform saccades to the right or left end of a horizontal bar that was presented at various locations on a video screen. A SEF neuron would fire strongly when the saccade was directed to the left end of the bar, but much more weakly when directed to the right end, irrespective of the location of the bar on the screen. About half of the SEF neurons showed object-related direction selectivity, each neuron favoring a particular location on the object, e.g., top, left end, bottom, right end, or center. In other words, SEF neurons encode the direction of an impending saccade relative to an object-centered frame of reference.

The "LIP–F7 dorsal" circuit may be important for controlling saccades that are made during head or body movements or, more generally, saccades concerned with complex motor programming (Pierrot-Deseilligny et al. 1995).

4.3.3.4
Ventrolateral Premotor Cortex (PMv; Areas F4 and F5)

Microstructure

The maps of the PMv proposed by different investigators are even more variable than those of the PMd. Not only do they differ in terms of location and size of the areas but also in their number: two regions were defined by von Bonin and Bailey (1947; areas FBA and FCBm) and Barbas and Pandya (1987; areas 6Va and 6Vb) whereas the Vogts (1919) proposed three areas: 6aα, 6bα, and 6bβ. Yet another area, namely, a rostral extension of the primary motor cortex, was introduced by the Vogts (1919; area 4c) and Barbas and Pandya (1987; area 4C). In both maps, this latter area is bordered medially by the arcuate spur and rostrally by the inferior arcuate sulcus.

Seen against this background, a multimodal approach based on different but complementary techniques yields more reliable and also (from a functional point of view; see below) more valid data. Based on a histochemical and cytoarchitectonic analysis, two areas have been proposed by Matelli et al. (1985): area F4 caudally and area F5 rostrally. More recently, receptor autoradiographic data (Matelli et al. 1996) and immunohistochemical data on the distribution of neurofilament proteins with antibody SMI-32 (Fig. 18; Geyer et al. 1998b; Matelli et al. 1996; Petrides et al. 2000) have confirmed this map. In Nissl-stained material, area F4 is poorly laminated (like areas F1, F2, and F3; Fig. 16C). Similar to area F2, pyramids increase in size from superficial to deep layer III, but in area F4 the overall cell density is lower. Scattered very large pyramidal cells (arrowheads in Fig. 16C) are present in its caudal (toward area F1) and especially dorsomedial part (toward the arcuate spur). Area F5 is clearly laminated with a prominent layer V (Fig. 16D). Cell density in area F5 is higher than in F4. Spur and inferior arcuate sulcus separate areas F4 and F5 from the PMd and the prefrontal cortex (Fig. 10C). Area F4 lies rostral to area F1 extending dorsomedially to the arcuate spur and ventrolaterally to the rostralmost extension of the primary somatosensory cortex at the level of the inferior precentral dimple. Area F5 occupies the rostral part of the PMv and abuts the prefrontal cortex in the fundus of the inferior arcuate sulcus. Preliminary evidence from SMI-32 immunohistochemistry indicates that (similar to areas F2d and F2v within area F2) a more fine-grained structural organization is evident within area F4 also: differences in layer V immuno-

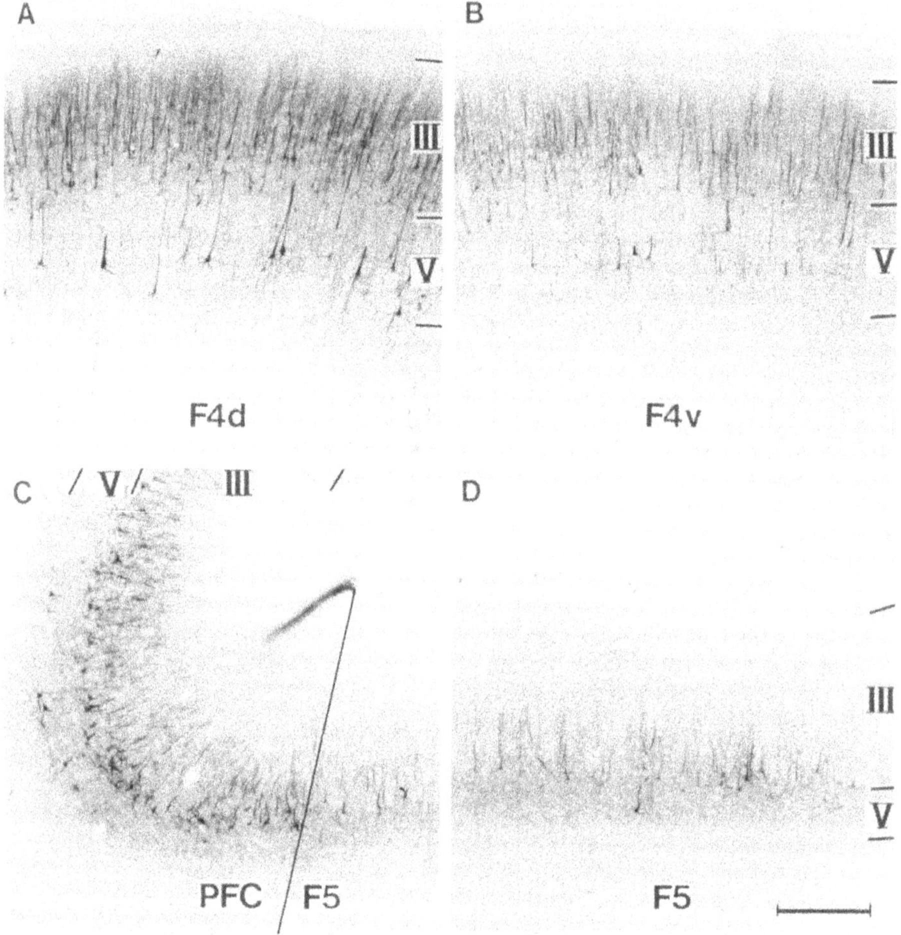

Fig. 18A–D SMI-32 immunoarchitecture of area *F4d* (**A**), *F4v* (**B**), the border between area *F5* and the prefrontal cortex (*PFC*; **C**), and area *F5* (**D**). Pyramidal cells in layers *III* and *V* are immunopositive. Differences in layer *V* immunoreactive pyramids define a dorsal (area *F4d*) and a ventral (area *F4v*) region within area F4. Immunoreactivity in area *F5* is very weak; a sharp increase marks the border between area F5 and the PFC. *Roman numerals* indicate cortical layers. Scale bar, 500 µm. (Reprinted from Geyer et al. 2000a)

reactive neurons define a dorsal (area F4d; Fig. 18A) and a ventral (area F4v; Fig. 18B) region within area F4 (Geyer et al. 1998b).

Connectivity

Area F4 receives a large thalamic projection from the nucleus ventralis lateralis pars oralis (VLo) and additional projections from area X of Olszewski (1952), nucleus ventralis posterolateralis pars oralis (VPLo), and nucleus ventralis lateralis pars caudalis (VLc). Area F5 receives its main thalamic input from area X. Additional projections to area F5 arise from the VPLo and VLc. Similar to areas F2, F3, F6, and F7, areas F4 and F5 are elements of different subcortical motor loops. Area F4 is the target (via nucleus VLo) of the putamen and pallidum, whereas area F5 receives its input (via area X) from the cerebellum (Matelli et al. 1989).

On a cortical level, area F5 (but not F4) receives afferents from the prefrontal cortex in the lateral bank of the principal sulcus. Area F4 is connected more strongly with area F3 than with area F6. Conversely, area F5 is connected more strongly with area F6 than with F3. Area F1 receives input from the entire area F4 but only from a small part of area F5 in the caudal bank of the inferior arcuate sulcus (Luppino et al. 1993; Matelli et al. 1986). Similar to areas F2 and F7, areas F4 and F5 receive rich cortical input from the posterior parietal cortex. These connections form several parieto-frontal circuits that are important functional units for the cortical control of reaching and grasping (Luppino et al. 1999; Rizzolatti et al. 1997, 1998). Area F4 receives its "predominant" parietal input from the ventral intraparietal (VIP) area in the fundus of the intraparietal sulcus (*"VIP-F4" circuit*; Figs. 10B, 11B). The anterior intraparietal (AIP) area in the lateral bank of the intraparietal sulcus projects mainly to the part of area F5 in the caudal bank of the inferior arcuate sulcus (*"AIP-F5 bank" circuit*; Figs. 10B, 11B), whereas area PF in the inferior parietal lobule projects predominantly to the sector of area F5 on the cortical convexity (*"PF-F5 convexity" circuit*; Figs. 10C, 11C).

Area F4 sends descending projections to the brain stem and spinal cord. The dorsal part of area F4 (arm representation; see below) projects to the brain stem reticular formation and the cervical spinal cord, the ventral part (face representation; see below) to the brain stem nuclei that control the orofacial muscles. Area F5 projects almost exclusively to the reticular formation except for a small sector in the caudal bank of the inferior arcuate sulcus that is also connected with the spinal cord (Dum and Strick 1991; He et al. 1993; Keizer and Kuypers 1989).

Function

The discussion of the function of areas F4 and F5 will also follow a more detailed description of the properties of these parieto-F4 and parieto-F5 loops.

"VIP-F4" Circuit. In contrast to the eye-centered representation of space found in area LIP, neurons in area VIP (Colby et al. 1993) in the fundus of the intraparietal sulcus code space with respect to a head-centered frame of reference. Neurons in area VIP are typically bimodal—visual and somatosensory; i.e., they respond to visual stimuli moving in the space within reaching distance around the body (peripersonal space; Colby et al. 1993) and also to light touch on the skin (Duhamel et al. 1991). The somatosensory receptive fields of these bimodal neurons are typically found on the head and face. Most interestingly, the visual and somatosensory fields of an individual VIP neuron match in location (a neuron that responds, e.g., to a visual stimulus in the upper right visual field also responds to touch in the upper right part of the face), size (a neuron with, e.g., a large visual receptive field also has a large somatosensory receptive field), and preferred direction (a neuron that responds, e.g., to a visual stimulus moving from left to right also responds to a somatosensory stimulus moving in the same direction across the skin; Fig. 19). This congruence is maintained when the eyes move, or, in other words, the visual receptive field of a given VIP neuron remains anchored to its somatosensory counterpart even when

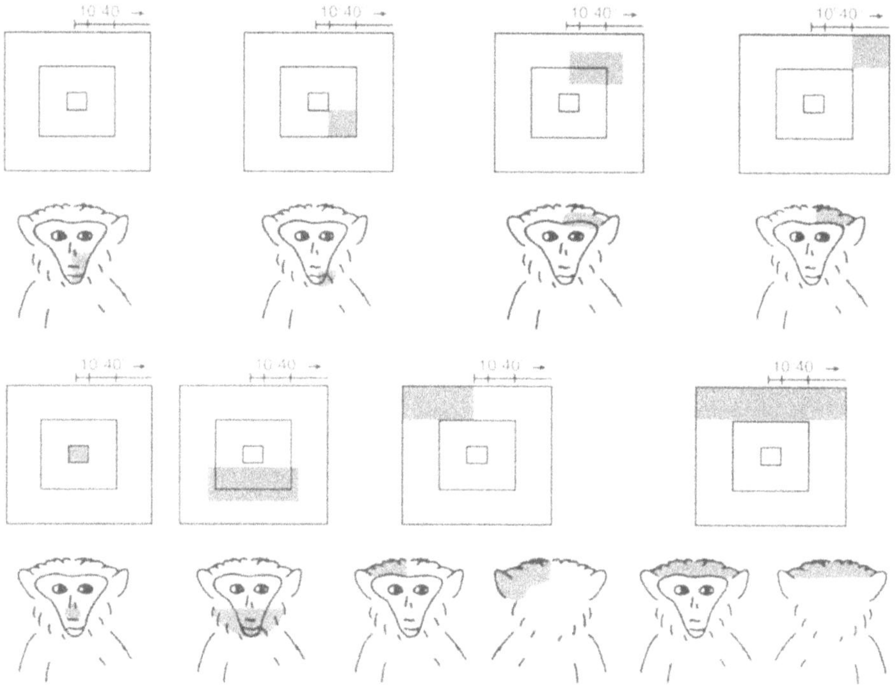

Fig. 19 Matching somatosensory and visual receptive fields of eight bimodal neurons recorded from macaque area VIP. On each outline of the monkey's head, the *gray patch* corresponds to one neuron's somatosensory receptive field. On the *square* above the head, the *gray rectangle* corresponds to the same neuron's visual receptive field. The *square* can be thought of as a screen in front of the monkey so that its center is directly ahead of the monkey's eyes. (Reprinted from Colby and Olson 1999 with persmission from Elsevier)

the stimulus activates a different set of retinal neurons. Thus, space is coded relative to a head-centered (and not eye-centered) frame of reference.

Area VIP projects predominantly to area F4. Area F4 is electrically excitable and organized in a somatotopic fashion: arm movements are represented dorsally and orofacial movements ventrally (Fig. 10C). ICMS shows that, in contrast to area F1, thresholds for eliciting movements are higher in area F4, evoked movements are more complex (involving more than two articulations or involving noncontiguous articulations or body parts, e.g., shoulder and wrist), and mainly proximal movements can be elicited from the arm field of this area (Gentilucci et al. 1988, 1989; Hepp-Reymond et al. 1994; Kurata and Tanji 1986). Similar to area VIP, many neurons in area F4 are also bimodal with somatosensory receptive fields on the face, neck, trunk or arms, and visual receptive fields in the peripersonal space and in register with their somatosensory counterparts (Fig. 20). Approaching objects are the most effective visual stimuli, whereby an increase in stimulus velocity typically expands the depth of the receptive field. In most cases, the visual receptive field does not move when the monkey moves the eyes. However, it does move when the monkey changes the position of the body part to which it is anchored (Fig. 20). Hence, in area

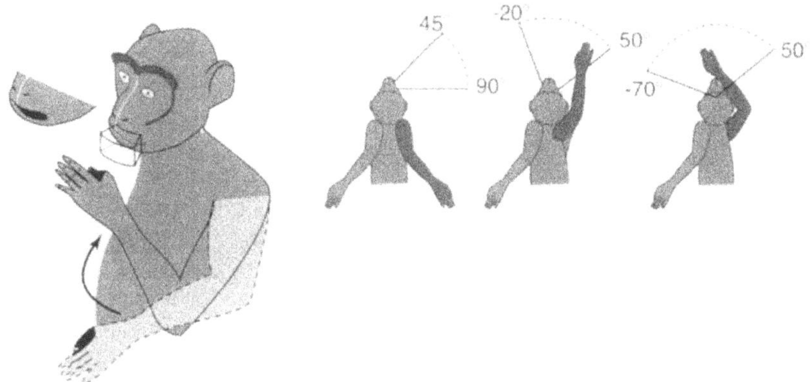

Fig. 20 *Left*: Matching somatosensory (perioral region) and visual (adjoining peripersonal space) receptive fields of one bimodal neuron recorded from macaque area F4. The same neuron responds also to tactile stimulation of the skin between thumb and index finger and to a goal-directed hand movement toward the mouth. *Right*: Bimodal F4 neuron with a somatosensory (right arm) and visual (peripersonal space in the face region) receptive field. The visual receptive field remains anchored to its somatosensory counterpart, i.e., it shifts from right to left as the arm is moved from right to left. (Reprinted from Schieber 1999 with permission from Elsevier)

F4, space is also coded in body-parts-centered coordinates (Fogassi et al. 1996; Gentilucci et al. 1988; Graziano et al. 1994). Many F4 neurons fire when reaching movements of the proximal arm are made, but not upon movements of the distal arm. Often a correlation is found between the positions of the somatosensory and visual receptive fields and the direction of the effective movement (a neuron with, e.g., a tactile and visual receptive field in the face region also fires during a movement toward the space above the shoulder; Fig. 20; Gentilucci et al. 1988).

These findings prompted Rizzolatti and colleagues (1998) to conclude that the "VIP–F4" circuit plays a role in encoding peripersonal space and in transforming object locations into appropriate movements toward them.

"AIP–F5 Bank" Circuit. Neural activity in area AIP (Sakata et al. 1995) in the lateral bank of the intraparietal sulcus was studied in monkeys trained to grasp different objects that required different patterns of hand and finger movements (pull lever with four fingers, push button with thumb, pull knob in a groove with tip of thumb and index finger, pull protruding knob with side grip of thumb and index finger) or simply to fixate these objects. The tasks were performed in light and darkness. Three different types of neurons can be differentiated in area AIP (Fig. 21). Motor-dominant neurons are active during object manipulation in light and darkness but not during object fixation. Most of them are more or less selective for one object. Visual-and-motor neurons have visual and motor properties. They are more active during object manipulation in light than in darkness or during object fixation. Visual-dominant neurons have purely visual properties. They respond to object manipulation in light and to object fixation but remain silent during object manipulation in darkness. An interesting finding is that the activity of neurons with motor properties is not influenced by the position of the object in space, which shows that these neurons en-

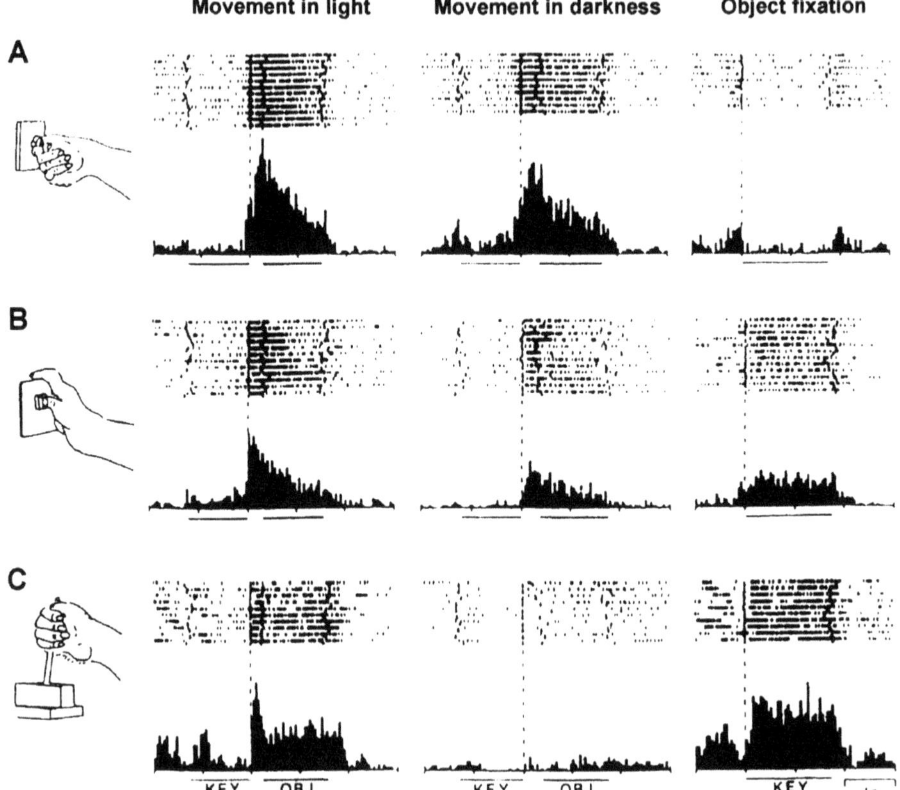

Fig. 21A–C Visual and motor response properties recorded from three different neurons (A–C) in macaque area AIP. A Motor-dominant neuron responds to object manipulation (pull protruding knob with thumb and index finger) in light and darkness, but not to object fixation alone. B Visual-and-motor neuron responds to object manipulation (push button with thumb) in light and darkness, but also to object fixation alone. C Visual-dominant neuron responds to visual stimuli (pull lever with four fingers in light and fixation of the same object without movement), but not to object manipulation in darkness. (Reprinted from Sakata 1996 with permission from Hideo Sakata)

code hand and finger but not proximal arm movements. Furthermore, neurons with motor and visual properties prefer the same object for fixation and manipulation, suggesting that these neurons match the intrinsic geometric properties of the object with the appropriate hand movement for grasping it. In a delayed-manipulation task, this information is even maintained for a short period after the object has disappeared (Murata et al. 1996; Sakata et al. 1995; Taira et al. 1990). When area AIP is reversibly inactivated by microinjecting muscimol, no major deficits in reaching are obvious, but the monkey is no longer able to appropriately preshape its hand and fingers in order to accommodate the object (Gallese et al. 1994). The results suggest that area AIP is crucially involved in the visual guidance of goal-directed hand movements.

Area AIP projects mainly to the part of area F5 in the caudal bank of the inferior arcuate sulcus. Area F5 is electrically much less excitable than area F4. Movements

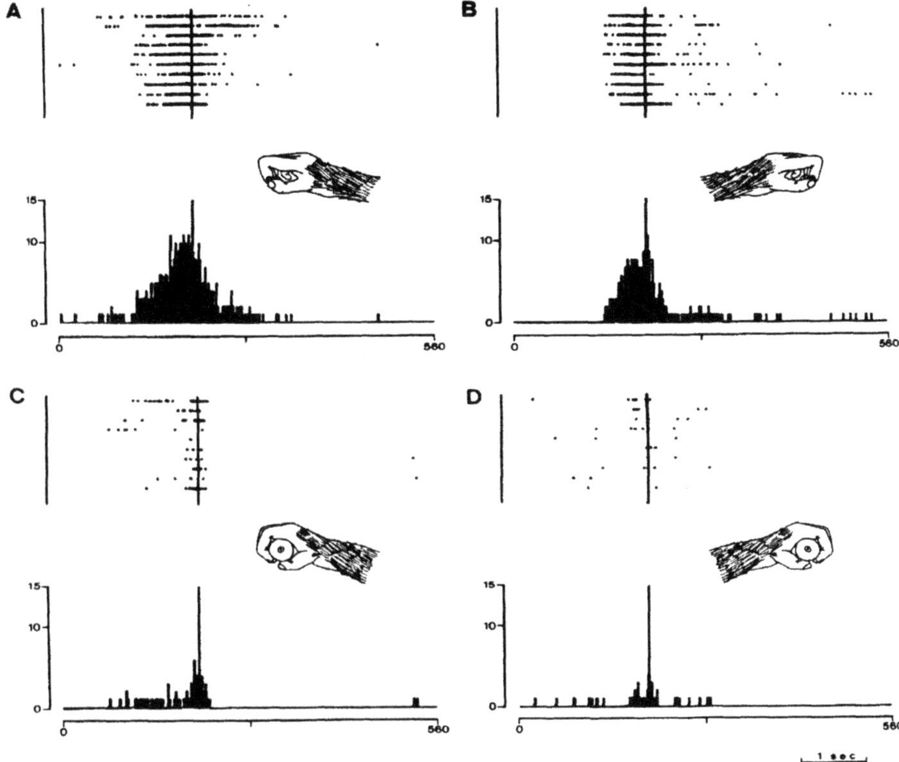

Fig. 22A–D Motor response properties recorded from one neuron in macaque area F5 during precision grip (**A**, **B**) and whole-hand prehension (**C**, **D**). The neuron was tested for contralateral (**A**, **C**) and ipsilateral (**B**, **D**) hand movements. The cell fired only upon precision grip both with the contra- and ipsilateral hand. (Reprinted from Rizzolatti et al. 1988 with permission from Springer)

can be elicited almost exclusively from the sector of area F5 lying in the caudal bank of the inferior arcuate sulcus. Area F5 is organized somatotopically. Similar to area F4, arm movements are represented dorsally and orofacial movements ventrally (Fig. 10C). In contrast to area F4, the distal arm is mainly represented in area F5 (whereas the proximal arm in area F4; Gentilucci et al. 1988, 1989; Hepp-Reymond et al. 1994; Kurata and Tanji 1986). Most area F5 neurons discharge not upon individual movements made by the animal but upon specific goal-directed motor acts, typically when grasping an object with a precision grip (i.e., opposition of the thumb and index finger), finger prehension (i.e., opposition of the thumb to the other fingers), or whole hand prehension (i.e., flexion of all fingers around the object; Fig. 22). When objects with different geometric properties are grasped in a similar way, similar neuronal responses are obtained. A fraction of these neurons also responds to somatosensory stimuli with a good correlation between the neuron's preferred type of grip and the location of the receptive field (e.g., a precision grip neuron has its somatosensory receptive field on the thumb and index finger). Some neurons also respond to visual stimuli, again with a good correlation between the neuron's preferred type of grip and the geometry of the object the cell is responding to (e.g., a precision grip neuron is also activated by a small visual stimulus but not by a large one; Rizzolatti et al. 1988). Even when a motor response to an object is not required, many F5 neurons discharge to its mere visual presentation (Murata et al. 1997). The motor deficits following reversible inactivation of area F5 are very similar to the symptoms upon inactivation of area AIP (see above): severe disturbance of hand preshaping and object grip, but no paralysis and no deficit in reaching (Gallese et al. 1997).

In sum, the "AIP–F5 bank" circuit is important for coding an object's intrinsic geometric properties and transforming them into appropriate goal-directed hand movements (Jeannerod et al. 1995). Or, to put it differently, each time an object is observed, its visual features are automatically (no matter whether a movement is intended or not) "translated" into a "vocabulary" of motor acts stored in area F5.

"PF–F5 Convexity" Circuit. Neurons located in the part of area F5 lying on the cortical convexity have the same motor properties as cells in the "F5 bank" region, i.e., they also discharge upon specific goal-directed motor actions. Only in the "F5 convexity" sector, however, can a group of neurons be found endowed with very intriguing visual properties as can be seen in the following paradigm. The monkey observes the experimenter grasping a small piece of food placed on a tray. The tray is then moved toward the monkey (into its peripersonal space) and the monkey itself grasps the food morsel. The neurons discharge when the experimenter grasps the food, stop firing when the food is moved toward the monkey, and discharge again when the monkey grasps the food. In other words, these neurons fire both when the animal performs a goal-directed motor action and when the animal observes another individual performing an action similar to that encoded by the neuron. Because of this correspondence between performing and observing an action, these neurons were termed "mirror neurons" (Gallese et al. 1996; Rizzolatti et al. 1996a). Mirror neurons, in order to be visually triggered, require an interaction between the agent of the action and the object. Observing the agent alone or the object alone (i.e., without interaction) are ineffective. The degree of correspondence between performing and observing an action can be very high. A typical mirror neuron fires when the monkey grasps an object and when the animal observes another individual grasping the same object. However, when the object is grasped with a tool (e.g., a forceps or a pair of pliers) instead of the hand, the neuron does not respond. On the other hand, the neuron does fire when the animal grasps the object in darkness. This indicates that the neuron indeed encodes a motor program and is not triggered by visual feedback.

How can this correspondence between acting and observing be explained? A straightforward interpretation would be that mirror neurons are related to motor preparation. A monkey that observes another monkey grasping food in a natural setting might implicitly prepare a similar action, e.g., in order to beat potential competitors for food. This explanation, however, is inappropriate since the neural activity decreases toward the end of observing the object and does not increase again until the monkey itself grasps the object. Were the activity indeed related to motor preparation, it should increase rather than decrease toward the end of the observation phase. A different interpretation is that mirror neurons are involved in the "understanding" of motor events (i.e., the ability to recognize that an individual is performing an action, to differentiate this action from other analogous to it, and to use this information in order to act appropriately; di Pellegrino et al. 1992). When an animal performs a specific motor act in a natural environment (e.g., grasping food), it knows from experience the consequences (in this case the emotionally positive feeling of being satiated). The motor action and its implicit meaning are encoded in a specific neuronal discharge pattern. Mirror neurons match this action (and its meaning) with the observation of the same action (and its meaning) in another individual. Performing an action (with its implicit meaning) and observing the same action result in the same neuronal discharge pattern. Hence, it may be assumed that observing an action in another individual also entails extracting (and thus "understanding") its biological meaning.

The "F5 convexity" region receives its main parietal input from area PF (Pandya and Seltzer 1982) in the rostral inferior parietal lobule (Figs. 10C, 11C). Preliminary observations have revealed that a similar observation/execution matching system also exists in area PF (Fogassi et al. 1998).

Taken together, the "PF–F5 convexity" circuit is a system that matches observation with execution of a motor action. This suggests an important cognitive role for these parts of the inferior parietal and lateral premotor cortex, to wit, internally representing and thus "understanding" motor events.

4.4
The Isocortical Motor System in Humans

The data on the motor system in macaques that have been described above are based on a very elementary biological principle: "What differs in structure (in this context: microstructure of the cerebral cortex) should also differ in function (and vice versa)." In other words, neurons with similar electrophysiological properties should lie within the same cortical area and, conversely, the properties of neurons should change across a microstructural border. This tenet can be tested in a straightforward and *direct* way in macaques, since upon completion of the electrophysiological experiments the brain can be sectioned, sections can be cell-stained, and penetration sites can be *directly* correlated with the microstructural (e.g., cytoarchitectonic) pattern. This correlation between cytoarchitecture and function has been demonstrated many times in macaques and has led to the very elaborate view of the isocortical motor system in this primate species.

Not so in humans. For obvious ethical reasons, functional in vivo and anatomical post-mortem studies cannot be performed in the same brain. This precludes a *direct* correlation of microstructure and function. As a consequence, investigators have sought *indirect* ways to achieve a match between structure and function.

One way is to start from anatomy. Several microstructural maps of the human frontal cortex have been published so far (Bailey and von Bonin 1951; Braak 1980; Brodmann 1909; Campbell 1905; Sarkissov et al. 1955; Smith 1907; Vogt and Vogt 1919; von Economo and Koskinas 1925). Especially Brodmann's (1909) map has been of tremendous impact over the last decades (and is still found in many neuroscience textbooks). However, these "classical" maps were published in print format. This puts them at a clear disadvantage when structural data from these maps are to be matched with functional data obtained from different brains. First, these maps are schematic drawings that reflect the topographical situation in *one* representative brain and do not address the problem of interindividual macro- (see, e.g., Ono et al.

1990) and microanatomical (see, e.g., Rademacher et al. 1993) variability. Second, these maps are "rigid" and are not based on a spatial reference system. They cannot be adapted to an individual brain and multimodal integration of structural and functional data is impossible. More recently published atlases, e.g., the reference system of Talairach and Tournoux (1988), are of limited value as well, as their cortical maps are not based on genuine microstructural data. Instead, the authors seem to have transferred each area from Brodmann's schematic drawing to a corresponding position on the cortex of their reference brain. In addition, Talairach and Tournoux give only the approximate position of an area (borders between areas are not indicated), nor do they address the problem of interindividual variability.

A complementary approach is to start from a functional point of view. Over the last years, noninvasive imaging techniques (e.g., PET or fMRI) have mapped the cerebral cortex with ever-increasing spatial resolution, but they relate foci of activation only to macroanatomical landmarks (i.e., gyri and sulci). Plenty of evidence in macaques and other nonhuman primates, however, has shown that it is microstructure that parallels function. Unfortunately, most microstructurally defined interareal borders in the human cortex do not match macroanatomical landmarks and they are topographically quite variable across different individuals (Amunts et al. 1999, 2000; Geyer et al. 1996, 1999; Rademacher et al. 1993; Rajkowska and Goldman-Rakic 1995; Roland et al. 1997; Roland and Zilles 1994, 1996; White et al. 1997). Hence, structural–functional correlations based only on macroanatomy are questionable and may account for at least some of the conflicting results functional imaging studies have provided in recent years, e.g., the debate whether (Hallett et al. 1994; Leonardo et al. 1995; Porro et al. 1996; Roth et al. 1996; Sabbah et al. 1995; Stephan et al. 1995) or not (Decety et al. 1994; Parsons et al. 1995; Rao et al. 1993; Roland et al. 1980; Sanes 1994) the human primary sensorimotor cortex is activated during imagined but not executed movements.

A third approach is to start from both anatomy and function and to fuse the two lines together. A recent development, the computerized brain atlas (Roland and Zilles 1994), offers the computational tools necessary to achieve this goal. On the one hand, genuine microstructural data (e.g., cytoarchitectonic analysis of cell-stained whole brain sections obtained from postmortem brains) are brought into the standard anatomical format of a computerized atlas. The degree of interindividual variability can be assessed by importing microstructural data from several brains (approximately ten in order to keep the time-consuming and cumbersome procedure of microstructural parcellation within reasonable limits). On the other hand, functional imaging data can be brought into the identical standard anatomical format. The two data sets can then be superimposed and correlated with each other on a probabilistic basis. This approach, termed *probabilistic microstructural–functional correlation* opens up the interesting possibility of (1) defining volumes of interest (VOIs) of motor areas that are not based on macroanatomical landmarks but instead on cytoarchitectonic mapping of postmortem brains and of (2) determining in these VOIs changes in regional cerebral blood flow data obtained from PET or fMRI experiments. In the human frontal cortex this new approach has been successfully used so far to map areas 4a and 4p of the primary motor cortex (Geyer et al. 1996) and areas 44 and 45 of Broca's speech region (Amunts et al. 1999) and to correlate microstructural VOIs of these regions with functional data (Amunts et al. 1998; Binkofski et al. 2000a, b,

2002; Bodegard et al. 2000; Ehrsson et al. 2000; Geyer et al. 1996; Indefrey et al. 1999; Naito et al. 1999, 2000).

Unfortunately, in humans, several anatomical techniques yield poorer results than in the macaque monkey (e.g., immunohistochemistry of neurofilament proteins with antibody SMI-32) or cannot be used altogether (e.g., in vivo tract-tracing techniques). Hence, a complete multimodal parcellation of the human isocortical motor system (similar to areas F1 to F7 in the macaque brain) is not yet available. As a consequence, there is still quite a way to go in humans before reaching a complete probabilistic microstructural map and, apart from the primary motor cortex, toward firmly establishing homologies between the human and macaque isocortical motor system. Nevertheless, some data do exist and some conclusions can be drawn.

4.4.1 Structural Organization

4.4.1.1 Primary Motor Cortex

Cytoarchitectonic features and topography of the human and macaque primary motor cortex are very similar. Low cell density, poor lamination, absence of layer IV, a diffuse border between layer VI and the underlying white matter, and giant pyramidal (Betz) cells in layer V also differentiate the primary motor cortex in humans (Brodmann's area 4; Fig. 23A, B) from the caudally adjoining primary somatosensory cortex. The border lies in the depth of the central sulcus close to its fundus. Scattered or absent giant pyramidal cells and large, elongated, and densely packed pyramids in lower layer III characterize the non-primary motor cortex (Brodmann's area 6; Fig. 23C) that rostrally abuts area 4. Dorso-medially (toward the midline), the rostral border of area 4 lies on the exposed cortical surface on the vertex of the precentral gyrus. Ventro-laterally (toward the Sylvian fissure), the rostral border of area 4 recedes in a caudal direction and eventually disappears in the depth of the central sulcus. The mesial part of area 4 occupies the paracentral lobule.

Recently, human area 4 has been subdivided into a caudal (area "4 posterior" or 4p; Fig. 23A) and a rostral (area "4 anterior" or 4a; Fig. 23B) region based on differences in cytoarchitecture and neurotransmitter binding sites (Geyer et al. 1996). The two areas are two parallel bands within area 4 running medio-laterally from the midline to the Sylvian fissure. Clear-cut changes in the laminar binding patterns, especially of muscarinic cholinergic (revealed with [^{3}H]oxotremorine-M) and serotoninergic (revealed with [^{3}H]ketanserine) binding sites, closely match corresponding cytoarchitectonic borders. Lower layer III pyramidal cells are small and loosely aggregated in area 4p (Fig. 23A), larger and more densely packed in area 4a (Fig. 23B), and even larger, more elongated, and sometimes arranged in several parallel rows like a phalanx in area 6 (Fig. 23C). There are no obvious differences in size, packing density or arrangement of giant pyramidal cells between areas 4a and 4p. Mean regional binding densities of many neurotransmitter binding sites tend to be higher in area 4p (more similar to the primary somatosensory cortex where binding densities are even higher; Geyer et al. 1997) whereas they are lower in area 4a (more similar to area 6 where binding densities are even lower). Hence, there is a clear neurochemical

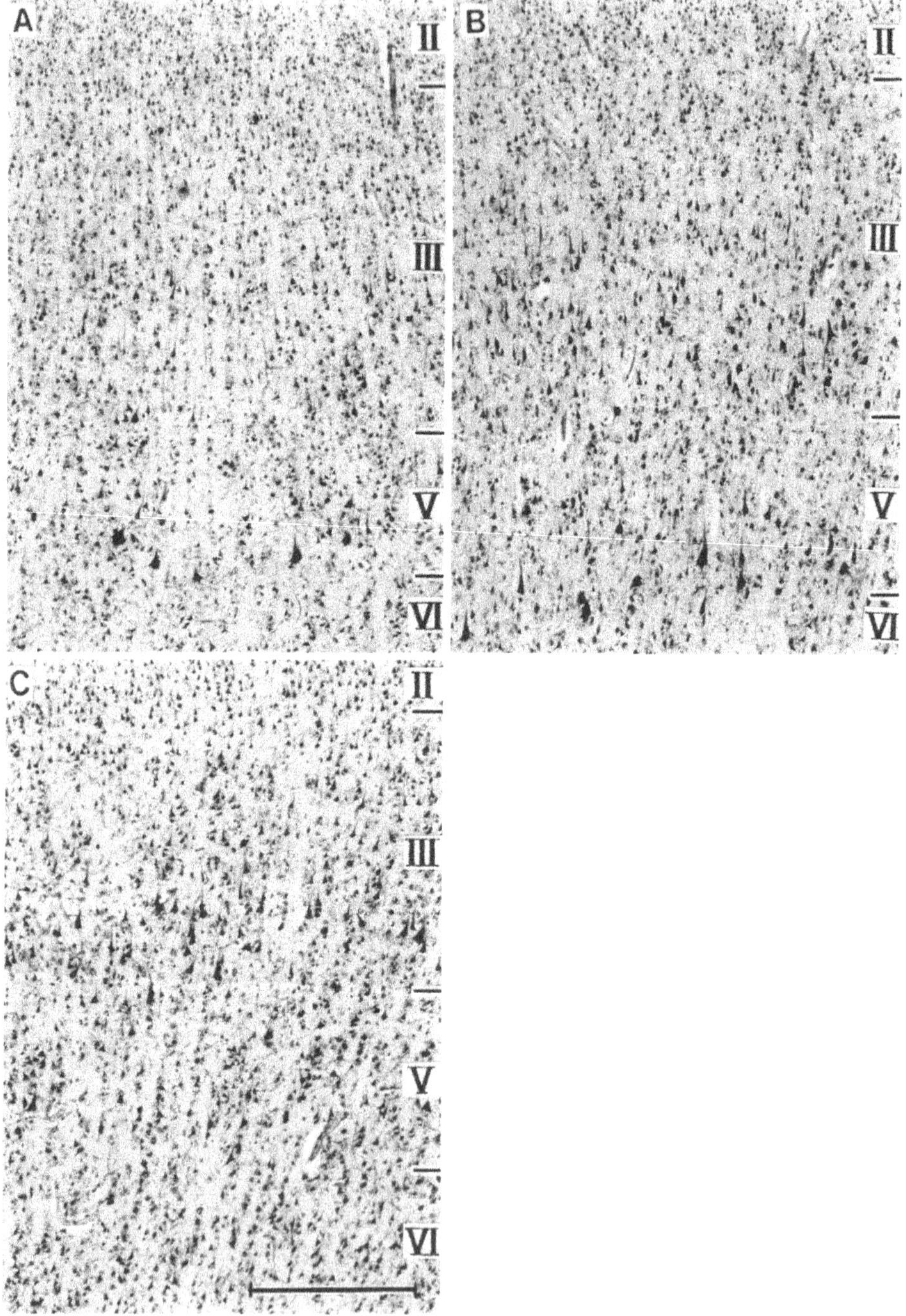

Fig. 23A–C Cytoarchitecture of human area 4p (**A**), 4a (**B**), and 6 (**C**). Lower layer *III* pyramids are small and loosely aggregated in area 4p, larger and more densely packed in area 4a, and even larger, more elongated, and sometimes arranged in several parallel rows in area 6. *Roman numerals* indicate cortical layers. Scale bar, 500 μm. (Reprinted from Geyer et al. 2000a)

dichotomy within area 4, the posterior part (area 4p) being more similar to the primary somatosensory cortex and the anterior part (area 4a) more similar to the non-primary motor cortex (area 6). At the moment, it is unclear whether there is any homology between human areas 4a and 4p and areas M1r and M1c recently described in the primary motor cortex of the New World owl monkey (*Aotus trivirgatus*) by Stepniewska and co-workers (1993; see above). The topography of areas 4a/4p and M1r/M1c is comparable in both species, but the cytoarchitectonic criteria based on which the two areas have been delineated in each species differ. In addition, should there be any homology (in terms of a common evolutionary ancestor), such a subdivision should also exist in the Old World macaque monkey [located in the evolutionary lineage between the hominoids (i.e., apes and humans) and the New World monkeys]. In macaques, however, no such subdivision has been found so far.

4.4.1.2
Supplementary Motor Areas "SMA Proper" and "Pre-SMA"

Cytoarchitectonic similarities between humans and macaques are also obvious on the mesial cortical surface. Two comparative studies (Zilles et al. 1995, 1996) found striking architectonic similarities between macaque areas F3 and F6 and the mesial parts of areas 6aα and 6aβ [according to the nomenclature of the Vogts (1919)], respectively, in humans. Area F3 and mesial area 6aα are characterized by increased cell density in lower layer III and layer V. Both areas are well demarcated on the cortical convexity from area F2 and from lateral area 6aα, respectively. Area F6 and mesial area 6aβ are clearly laminated and are characterized by a prominent layer V well separated from layers III and VI. Both areas are well demarcated on the cortical convexity from area F7 and from lateral area 6aβ, respectively. Additional similarities have been found on a neurochemical basis (Geyer et al. 1998a; Zilles et al. 1995, 1996). For example, [^{3}H]kainate binding sites are mainly concentrated in the deep layers in both areas in macaques and humans. Mean binding density in area F3/mesial 6aα is higher than in the primary motor cortex and lateral premotor cortex in both species. Binding sites of [^{3}H]oxotremorine-M show a superficial and deep cortical band of maximal binding densities in both species. Mean binding densities increase in a caudorostral direction from the primary motor cortex to area F6 in macaques and to mesial area 6aβ in humans. The authors concluded that human mesial areas 6aα and 6aβ are possibly homologous to macaque areas F3 (SMA proper) and F6 (pre-SMA), respectively.

In humans, the border between the primary motor cortex and mesial area 6aα (possibly SMA proper) coincides approximately with the VPC line (cf. Fig. 9A). The border between mesial area 6aα and mesial area 6aβ (possibly pre-SMA) coincides approximately with the VAC line (Zilles et al. 1996). The coincidence of the borders between the primary motor cortex/SMA proper and the SMA proper/pre-SMA with the VPC and VAC line, respectively, has also been shown by Vorobiev et al. (1998). In the latter study, human SMA proper has been further subdivided into a caudal (SMAc) and a rostral (SMAr) part. Whether these two subdivisions reflect somatotopy or a more fine-grained functional differentiation within human SMA proper is an unresolved issue at the moment.

In a recent study (Grosbras et al. 1999) a macroanatomical landmark was sought that indicates the position of the human supplementary eye field (SEF). Such a landmark was indeed found, namely, the upper part of the paracentral sulcus on the mesial cortical surface "in the anterior part of the region usually described as the SMA-proper, and posterior to the VAC line, which is usually considered as being the posterior limit of the pre-SMA" (Grosbras et al. 1999, p. 710). This position of the human SEF on the mesial cortical surface is at a clear variance with the position of its monkey homologue on the dorsolateral convexity in area F7 (see above). There is no straightforward way to explain this discrepancy. Some clues may be gained from a functional imaging study (Fink et al. 1997) that detected an arm and a leg representation in the human mesial non-primary motor cortex that was confined to the *caudal* half of the region between the VPC and VAC line, and that could correspond to area SMAc. Macaque area F3 (SMA proper) contains a representation of the arm, leg, and face, the latter not tested in the imaging study. Hence, it might be that only area SMAc is the human homologue of macaque area F3 and area SMAr contains the human SEF that (perhaps due to an evolutionary shift) has moved from a dorsolateral position in the macaque to a mesial position in the human cortex.

4.4.1.3
Dorsolateral and Ventrolateral Premotor Cortex

On the cortical convexity of the macaque, a conspicuous macroanatomical landmark, the arcuate sulcus, separates the agranular frontal from the granular prefrontal cortex and the border between them looks like a horizontally mirrored "c" (Fig. 24A). A comparable sulcus does not exist in the human brain, but the spatial distribution of the agranular cortex (Brodmann's areas 4 and 6 plus dysgranular area 44; see below for a further discussion of areas 44 and 45) is strikingly similar. Its rostral border also bears some resemblance to a horizontally mirrored "c" (Fig. 24B). In both species the prefrontal cortex extends caudally in the middle of the dorsolateral convexity as if pushing back the agranular cortex, and, conversely, recedes rostrally in the dorsomedial and ventrolateral parts of the convexity.

In a recent review article (Rizzolatti et al. 1998) an attempt was made to define homologies between the human and macaque dorso- and ventrolateral premotor cortex. Macaque data were compared with data obtained in humans from cytoarchitectonic parcellation (Fig. 25D; Vogt and Vogt 1919), motor representations as determined with electrical stimulation (Fig. 25D; Foerster 1936), sulcal ontogeny (Turner 1948), and the putative location of the human frontal eye field in functional imaging studies (Paus 1996). It was proposed that the superior frontal and superior precentral sulcus (black in Fig. 25E) represent the human homologue of the macaque superior arcuate sulcus (black in Fig. 25C). Accordingly, the two areas that occupy the rostral part of the precentral gyrus and the caudal part of the superior frontal gyrus (superior part of area 6aα and area 6aβ according to Vogt and Vogt 1919) correspond to macaque areas F2 and F7, respectively. The inferior frontal sulcus and the ascending branch of the inferior precentral sulcus (dark gray in Fig. 25E) correspond to the macaque inferior arcuate sulcus (dark gray in Fig. 25C). The descending branch of the inferior precentral sulcus (black in Fig. 25E) corresponds to the inferior precentral dimple of the macaque monkey (black in Fig. 25C). Accordingly, the inferior part

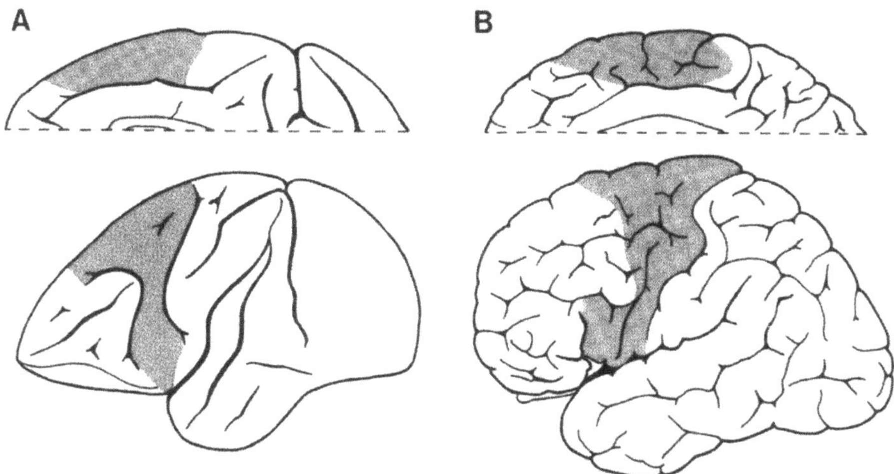

Fig. 24A, B Topography of the agranular frontal cortex (*dark gray*) in the brain of *Cercopithecus* (**A**), an Old World monkey closely related to the macaque, and in *Homo sapiens* (**B**). *Top part* of each figure shows the dorsal part of the mesial cortical surface, *bottom part* shows the lateral surface. In both species, the border between the agranular frontal and the rostrally adjoining granular prefrontal cortex resembles a horizontally mirrored "c". (Maps redrawn from Brodmann 1909. Reprinted from Geyer et al. 2000a)

of area 6aα (according to Vogt and Vogt 1919) and area 44 (according to Brodmann 1909) should be homologous to macaque areas F4 and F5, respectively. An open issue is the macaque homologue of human area 45 (according to Brodmann 1909). An area 45 was described in the macaque by Walker (1940) lying in the rostral bank of the inferior arcuate sulcus and related to eye movements (Bruce et al. 1985; Suzuki and Azuma 1983). Human area 45, however, is (together with area 44) the cytoarchitectonic correlate of Broca's speech region (Aboitiz and Garcia 1997) and there are no reports in the literature that human area 45 might be involved in eye movements. On the other hand, when one considers the cytoarchitectonic similarities between human areas 44 and 45 and their similar times of onset of myelinization (cf. Vogt and Vogt 1919), an alternative interpretation would be that human areas 44 and 45 evolved from one and the same area in the macaque monkey, namely, area F5.

4.4.2 Functional Organization

Functional evidence from neuroimaging studies indicates that the human motor cortex shares several general organizational principles with the macaque motor cortex. One important aspect is the existence of multiple representations of body movements in humans also. For example, simple flexion and extension movements of the fingers induce several independent activations in the primary motor cortex, SMA proper, dorsal premotor cortex, frontal operculum, and the cingulate motor areas (Fink et al. 1997). In addition to the primary (SI) and secondary (SII) somatosensory

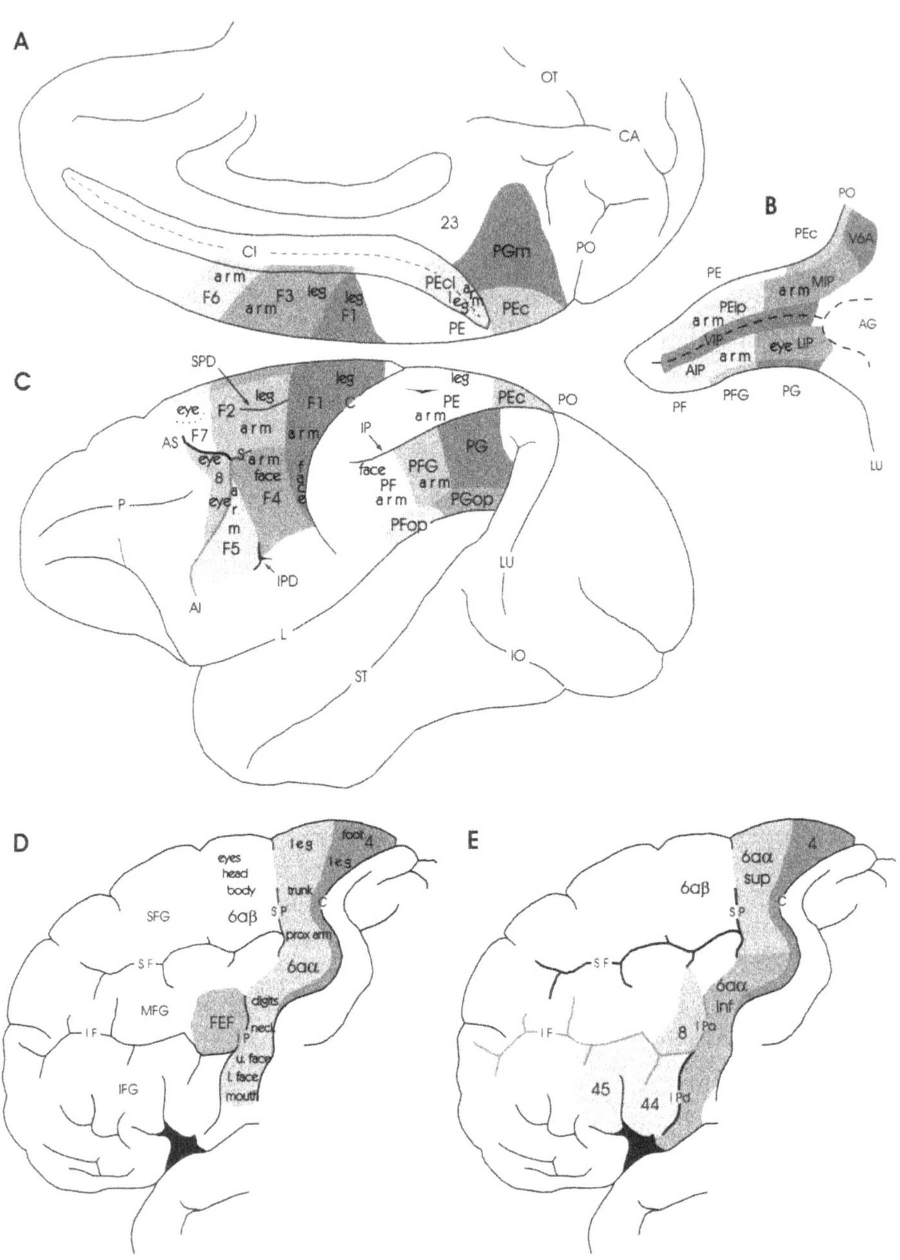

A
OT
CA
23
PGm
PO
CI
arm
F6
F3
leg
arm
leg
F1
PEcl
PEc
PE
B
PO
PEc
V6A
PE
arm
MIP
PEip
arm
VIP
AG
eye
LIP
AIP
arm
PF
PFG
PG
LU
C
SPD
leg
F1
leg
leg
PE
PEc
PO
eye
F2
C
arm
F7
arm
arm
IP
AS
arm
PG
eye
8
face
F4
face
PFG
eye
PF
arm
arm
PGop
P
PFop
m
F5
IPD
LU
AI
L
IO
ST
D
eyes
head
body
leg
foot
4
leg
trunk
SFG
$6a\beta$
SP
C
prox arm
SF
$6a\alpha$
MFG
FEF
digits
neck
IF
IP
u. face
l. face
IFG
mouth
E
4
$6a\alpha$
sup
$6a\beta$
SP
C
SF
$6a\alpha$
inf
IF
8
IPa
45
44
IPd

area, two further arm and finger representations are located in the superior and inferior parietal lobule (Fink et al. 1997). These observations fit well with the notion that multiple motor representations are also present in the posterior parietal cortex. Further evidence in favor of the existence of multiple representations of body movements in the frontal and posterior parietal cortex comes from studies of motor imagery (see Jeannerod and Decety 1995). A recent study in which motor imagery was compared with motor execution suggests that largely overlapping neural substrates in the frontal and parietal cortex are involved in both aspects of motor control (Gerardin et al. 2000). However, an open question is whether or not the primary motor cortex is involved in motor imagery. Several studies have shown that the precentral gyrus is activated during imagined movements, although to a lesser extent than during executed movements (Gerardin et al. 2000; Porro et al. 1996; Roth et al. 1996). Since these studies related the activations to macroanatomical landmarks (precentral gyrus) and not to cortical areas (4 and 6), one can only speculate as to what extent the primary motor cortex (area 4) has been involved. The activation of a network of parietal and frontal areas during motor imagery and execution suggests that the parieto-frontal circuits described above in macaques are key functional units also in humans. With experimental paradigms comparable to those in macaques, their topography in humans could recently be elucidated in greater detail. Binkofski et al. (1999) asked subjects to manipulate with their hands and fingers complex 3-D objects. They found activations in the intraparietal sulcus (possibly the human homologue of macaque area AIP) and the ventrolateral premotor cortex (possibly area 44 due to an activation of the human homologue of the "AIP–F5 bank" circuit). Bremmer et al. (2001) stimulated subjects with moving visual, tactile, and auditory stimuli and found activations in all three modalities in the depth of the intraparietal sulcus (possibly the human homologue of macaque area VIP) and the ventrolateral premotor cortex (possibly the human homologue of macaque area F4 due to an activation of the "VIP–F4" circuit).

The final sections provide an overview of the functional organization of the isocortical motor units in humans (i.e., primary motor cortex, supplementary motor areas "SMA proper" and "pre-SMA," dorsolateral premotor cortex, and ventrolateral premotor cortex).

Fig. 25 Mesial (**A**) and lateral (**C**) view of the macaque brain and unfolded view of the intraparietal sulcus (**B**). For further details see Fig. 10. **D, E** Lateral view of the human brain with a map of the agranular frontal cortex and the somatotopic representations of the body according to the Vogts (1919) and Foerster (1936; **D**) and the suggested homologies with the macaque cortex (**E**). *Identical gray values* in **C** and **E** indicate areas and sulci considered to be homologous. See text for further details. Numerals *4, 8, 44, 45* are areas according to Brodmann's nomenclature; *C*, central sulcus; *FEF*, frontal eye field; *IF*, inferior frontal sulcus; *IFG*, inferior frontal gyrus; *IP*, inferior precentral sulcus; *IPa*, inferior precentral sulcus (ascending branch); *IPd*, inferior precentral sulcus (descending branch); *MFG*, middle frontal gyrus; *SF*, superior frontal sulcus; *SFG*, superior frontal gyrus; SP, superior precentral sulcus. (Reprinted from Rizzolatti et al. 1998 with permission from Elsevier)

4.4.2.1
Primary Motor Cortex

The most complete functional map of the human precentral cortex was provided in the 1930s to 1950s by Penfield (Penfield and Boldrey 1937; Penfield and Rasmussen 1952), who stimulated the cortex of epileptic patients with surface electrodes during surgery. The results, summarized in the famous cartoon of the "motor homunculus," showed a somatotopic organization of the human precentral region that replicated Woolsey's findings in nonhuman primates (cf. "motor simiusculus" in the monkey). In this respect it is important to note that surface electrical stimulation reveals mainly the cortical origin of the descending corticospinal tract which in the macaque is not restricted to the primary motor cortex (area 4). When considering that the free surface of the precentral gyrus in humans is mostly occupied by area 6 (especially close to the Sylvian fissure; cf. Fig. 9D), it is reasonable to conclude that a significant part of Penfield's homunculus belongs to the non-primary motor cortex (area 6). The somatotopic organization of the precentral region has been replicated in recent years with functional imaging techniques (e.g., PET) in individual subjects (Fink et al. 1997). Despite an orderly representation of the body's periphery in each individual subject, topographical variability has been detected between individuals (Fink et al. 1997). Nonetheless, a macroanatomical landmark exists that reliably indicates the hand area, to wit, a knob-like structure in the precentral gyrus that looks like an omega or epsilon in the axial plane and that corresponds to the "middle knee" of the central sulcus in postmortem brains (Yousry et al. 1997). As in macaques, somatotopic organization should not be overinterpreted. Individual finger movements are also represented in humans as multiple and spatially overlapping activation sites (Sanes et al. 1995).

Activations in the caudal bank of the precentral gyrus (or rostral bank of the central sulcus) have been detected in many PET and fMRI studies with a wide variety of different motor paradigms. Since the foci of activation were related to macroanatomical landmarks and not to cortical areas, none of these studies could exclude whether the foci extended (at least to some extent) into the rostrally adjoining area 6. The technique of microstructural–functional correlation with a computerized brain atlas (see above) overcomes this dilemma. In a recent study, PET foci were correlated with the population maps of primary motor areas 4a and 4p (Geyer et al. 1996). Roughness discrimination of two cylinders with different microprofiles with the right thumb and index finger activated area 4p significantly more than did a control condition of self-generated movements without object interaction. Neurochemically, area 4p is more similar to the primary somatosensory cortex whereas area 4a is more similar to the non-primary motor cortex (area 6; Geyer et al. 1996). This difference in neurochemistry might be reflected on a functional level in such a way that a voluntary motor act that is closely modulated by somatosensory feedback (i.e., scanning the surface texture of an object) leads to a stronger activation of area 4p.

4.4.2.2
Supplementary Motor Areas "SMA Proper" and "Pre-SMA"

The role of the SMA in human motor control has been a matter of debate for a long time. Electrical stimulation studies, clinical observations of deficits caused by vascular lesions or surgical ablations, recordings of cortical potentials during the execution of motor tasks, as well as early functional imaging studies provided conflicting results (for a review of the literature see Freund 1996). On the one hand, the high electrical excitability of the SMA and its activation during the execution of simple aimless movements were taken as evidence for an "executive" role in motor control. On the other hand, the presence of global akinesia without paralysis, the observation of slow negative potential shifts long before the execution of a motor task, as well as its activation when subjects were requested to mentally rehearse a series of digit movements led to the concept of a "supramotor" function of the SMA. However, in all these studies no attempt was made to precisely identify the anatomical location of the stimulated or damaged cortex, mainly because of the assumption that the SMA is a structurally and functionally homogeneous entity. The discovery, in nonhuman primates, that the mesial agranular cortex consists of two different structural and functional areas led to a re-evaluation of these apparently conflicting data in humans.

As mentioned above, there is increasing evidence that the human SMA also consists of two areas, namely, SMA proper and pre-SMA. The border between the two areas does not coincide with any macroanatomical landmark; it is thought to approximately correspond with the VAC line. Due to similarities in cytoarchitecture and neurochemistry, human SMA proper and pre-SMA are possibly homologous to macaque areas F3 and F6, respectively (Zilles et al. 1995, 1996).

When functional data are correlated with the location of the VAC line, it becomes obvious that the two areas caudal and rostral to it are also functionally distinct and their roles in motor control are in many ways comparable to those proposed for macaque area F3 (SMA proper) and F6 (pre-SMA). The results of electrical stimulation studies of the human mesial frontal cortex at first glance do not seem to be consistent with this proposed homology, since stimulating cortical sites on either side of the VAC line evokes relatively low-threshold movements. However, by using chronically implanted subdural electrodes and better stimulation parameters, a somatotopic organization of the SMA proper comparable to the macaque could be demonstrated in humans also (Fried 1996; Lim et al. 1996). In contrast, electrical stimulation of a region spanning about 3 cm rostral to the region where movements can be elicited more frequently (i.e., SMA proper) induced "negative symptoms" (e.g., transient speech arrest) or more complex effects (e.g., an "urge" to move or the feeling of an impending movement; Fried 1996). Such results led Fried to consider the caudal "motor excitable" zone to be comparable with the macaque SMA proper (area F3) and the rostral "negative motor" or "motor arrest" zone to be comparable to macaque pre-SMA (area F6).

In a review article of the neuroimaging literature, Picard and Strick (1996) studied activations in the mesial frontal cortex depending on the complexity of movements. The authors classified as "simple" those tasks that required the most basic spatial or temporal organization of movements, or tasks that were overlearned and highly practiced (e.g., moving a joystick in a fixed direction upon an auditory trigger signal). "Complex" tasks were characterized by additional motor or cognitive demands such as the selection of a motor response (e.g., moving a joystick in a random, self-selected direction upon an auditory signal) or the acquisition of a conditional association that determined the motor response. "Simple" tasks typically activated regions caudal, "complex" tasks rostral to the VAC line (Picard and Strick 1996).

Another important aspect is the execution of motor sequences. Functional data indicate that SMA proper and pre-SMA are differently involved in this type of motor task: SMA proper appears to play a role in the execution of well-learned, "automatic" motor sequences, whereas pre-SMA seems to be crucial for learning new, "unfamiliar" sequences. Sergent et al. (1992) impressively demonstrated this difference in a study with professional pianists. SMA proper was activated when subjects were playing scales (i.e., a highly automatic movement sequence), whereas pre-SMA became active when the same subjects were playing an unfamiliar piece of music. When subjects were studied at different stages of learning motor sequences, the activation shifted from pre-SMA to SMA proper as the task became more routine and eventually "automatic" (Grafton et al. 1992a; Jenkins et al. 1994; Schlaug et al. 1994; Seitz and Roland 1992). Finally, Hikosaka et al. (1999) showed that pre-SMA was involved in learning motor sequences by trial and error, while executing a learned sequence activated SMA proper.

A more cognitive role of pre-SMA was suggested in a fMRI study by Petit et al. (1998) in which sustained activity in pre-SMA was observed during the delay period of both spatial and visual working memory tasks. The authors concluded that this activity may reflect an involvement of pre-SMA in cognitive processes higher than those necessary for selecting or preparing a movement. This proposed role of the human pre-SMA as a "supramotor" control center is very similar to that proposed for monkey area F6 (cf. Rizzolatti et al. 1996c, 1998).

Finally, a set of experimental data sheds light on the role of SMA proper in motor imagery. These data suggest a possible functional subdivision of SMA proper that fits well with the recently proposed cytoarchitectonic subdivision of SMA proper into a caudal (SMAc) and a rostral (SMAr) part (Vorobiev et al. 1998). Although the physiological basis of motor imagery is still poorly understood, it was suggested that this process may be functionally similar to motor preparation and that imagined movements may share common anatomical and functional substrates with actual movements (Jeannerod 1994). Functional studies of imagined and executed arm movements showed two distinct activation foci, both located behind the VAC line: a more caudal one (possibly in area SMAc) related to motor execution and a more rostral one (possibly in area SMAr) related to motor imagery (Roth et al. 1996; Stephan et al. 1995; Tyszka et al. 1994). In a study by Grafton et al. (1996a) subjects were tested while observing actions of other individuals and imagining themselves performing them. In both conditions, only SMAr was activated.

Taken together, these data show that striking similarities do exist between SMA proper and pre-SMA in macaques and two possibly homologous areas in humans. In contrast to macaques, a greater degree of structural and functional fractionation of the mesial agranular cortex may have emerged during human evolution.

4.4.2.3
Dorsolateral Premotor Cortex

Electrophysiological, clinical, and neuroimaging evidence suggests that there are similarities also between the human dorsolateral premotor cortex and its presumed homologue in the monkey. Stimulation studies in humans showed that the electrically excitable cortex extends approximately 2–3 cm rostral from the central sulcus into the caudal sector of area 6 (Uematsu et al. 1992). This finding suggests that, as in the monkey, the caudal region of the human dorsolateral premotor cortex is the origin of a corticospinal projection. Lesion studies pointed out an involvement of the dorsolateral premotor cortex in both preparation/execution of motor acts and motor learning. Unilateral damage of the dorsolateral premotor cortex caused an inability to perform arm or leg movements requiring temporal coordination of proximal muscles, whereas the performance of distal skilled movements was normal (Freund and Hummelsheim 1985). PET data provided further evidence for an involvement of the dorsolateral premotor cortex in the motor control of arm and leg movements (Fink et al. 1997). As for arm movements, Colebatch et al. (1991) found an activation in the superior frontal gyrus associated with the execution of simple movements involving proximal and distal joints. The mean regional cerebral blood flow (rCBF) increase associated with shoulder movements was significantly higher than with distal movements suggesting that in this region, as in the monkey dorsolateral premotor cortex, there is an extensive representation of proximal movements. In agreement with these data, several neuroimaging studies showed that this cortical region was activated during tasks involving proximal and/or distal arm movements (Deiber et al. 1991; Grafton et al. 1992b, 1996b; Seitz et al. 1997). As in the monkey, the human dorsolateral premotor cortex is involved in movement preparation (Kawashima et al. 1994) and in movement selection according to arbitrary rules. Learning complicated finger movement sequences (Seitz and Roland 1992), acquiring new key press sequences on the basis of trial and error (Jenkins et al. 1994), and performing reaching and pointing movements according to the internal representation of targets (Kawashima et al. 1995) all activated the superior frontal gyrus. Involvement of the human dorsolateral premotor cortex in movement selection according to arbitrary rules was demonstrated in two PET studies: activations were detected as subjects learned to make an association between a cue and the direction of a movement (Deiber et al. 1997) and when subjects had to select more complex movements (e.g., precision or power grip) based on arbitrary cues (Grafton et al. 1998). This particular role of the dorsolateral premotor cortex had already been identified in an earlier clinical study showing that patients with lesions of this cortical region (but *not* of parietal or primary motor cortex) failed to associate hand movements with arbitrary sensory cues (Halsband and Freund 1990). A possible functional segregation within the dorsolateral premotor cortex was proposed by Iacoboni et al. (1998). In a spatial compatibility task, the authors found an activation more rostral in the dorsolateral premotor cortex when, in both the auditory and visual modality, the incompatible and compatible conditions were compared. In the same experiment, learning-related activation was observed in a more caudal part. It was suggested that these two different parts of the human dorsolateral premotor cortex could correspond to the rostrocaudal subdivision into areas F7 and F2 of the macaque monkey.

4.4.2.4
Ventrolateral Premotor Cortex

Functional studies of the ventrolateral premotor cortex initially focused on the role of Broca's area in speech functions. Only recently, neuroimaging studies showed that the human ventrolateral premotor cortex is also involved in the control of hand movements. These observations strongly support the proposed homology between Broca's area in humans and area F5 in macaques.

Early neuroimaging studies of subjects performing grasp movements failed to identify activations in the ventrolateral premotor cortex. First evidence for a hand representation in this region came from studies of learning finger movement sequences (Seitz and Roland 1992), mental imagery of grasping (Decety et al. 1994; Grafton et al. 1996a), and preparation of finger movements on the basis of copied movements (Krams et al. 1998). More direct evidence that the ventrolateral premotor cortex contains a hand representation came from a recent study of Binkofski et al. (1999). The authors found that manipulating complex objects with the hand and fingers resulted in an activation of presumed area 44 while covertly naming the manipulated objects led to an additional activation more rostrally in presumed area 45. One very interesting result of this study was that manipulating objects also activated parietal areas, one of them lying in the anterior part of the lateral bank of the intraparietal sulcus (possibly the human homologue of macaque area AIP). It seems that the parieto-frontal circuits for object manipulation described in macaques also exist in humans, and the data of Binkofski et al. (1999) suggest an activation of the human homologue of the "AIP–F5 bank" circuit.

Several lines of evidence indicate that the action recognition or "mirror" system described in macaque area F5 also exists in the human brain. First, experiments with transcranial magnetic stimulation showed that observation of hand actions facilitated the motor cortex of the observer, as shown by an increase in motor evoked potential amplitude in those hand muscles of the observer that matched the activated muscles in the observed hand (Fadiga et al. 1995; Strafella and Paus 2000). Second, electrophysiological studies demonstrated that observing hand actions desynchronized the motor cortex of the observer similar to actually performing hand actions (Cochin et al. 1999; Hari et al. 1998). Finally, brain imaging studies showed that observing hand actions activated a region in the ventrolateral premotor cortex that seemed to correspond to Broca's area (Decety et al. 1997; Grafton et al. 1996a; Grezes et al. 1998; Iacoboni et al. 1999; Rizzolatti et al. 1996b). The possibility that the activation of Broca's area during observation of action is due to internal verbalization rather than a mirror mechanism was ruled out in a recent study of Buccino et al. (2001).

Taken together, these data suggest that in Broca's area (as in macaque area F5) there is a mouth representation but also a representation of hand movements. At first glance, it may seem surprising that this cortical region—despite its pre-eminent role as a motor speech area in humans—is also important for internally representing and thus "understanding" motor events. It is beyond the scope of this book to discuss theories of human language evolution that are based on this "mirror neuron" concept. The results, however, support the theory that Broca's area in humans and area F5 in macaques are indeed homologous entities.

5 Epilogue

More than a century after Jackson's revolutionary concept that the human motor system is organized in a hierarchical way with the motor cortex representing the top level, one may observe that the data accumulated ever since still support this view. Although our knowledge of the detailed organization of the isocortical motor system is still very fragmentary, the definition of parieto-frontal circuits suggests that, in addition to a hierarchical component, there are also parallel pathways that influence cortical motor control.

Over the last few years, numerous studies have provided new insights into the structural and functional organization of the human cortical motor system. The data reviewed in this book indicate that striking similarities have been found between humans and nonhuman primates. In contrast to laboratory animals, however, ethical constraints greatly limit the use of invasive techniques in humans. This is why, in contrast to macaques, our knowledge of the human cortical motor system is still fragmentary.

During the first half of the twentieth century, several morphologists, among them Cécile and Oskar Vogt, proposed maps of the human motor cortex that are more congruent with our current knowledge than is Brodmann's "classical" parcellation. Nevertheless, Brodmann's map is still the "gold standard" for correlations between structure and function in neuroimaging studies.

Hence, it will be extremely important to establish in the near future a more detailed architectonic map of the human agranular frontal cortex compatible with areas F1–F7 of the macaque monkey. From a technical point of view, the stage for this endeavor has been set over the last several years. From a biological point of view, the first step has been accomplished in this study.

Acknowledgements. I would like to express my sincere thanks to Prof. Massimo Matelli, Giuseppe Luppino, and Giacomo Rizzolatti (Istituto di Fisiologia Umana, Università di Parma). During a 16 months' research stay from April 1996 through July 1997 in the serene ambience of the Emilia Romagna we launched the project outlined in this book.

Thanks are due to Prof. Per E. Roland (Karolinska Institute, Stockholm) for a long and fruitful collaboration that culminated in the implementation of the computerized brain atlas, to Dipl. Phys. Hartmut Mohlberg (Institute of Medicine, Research Center Jülich) for refining the spatial normalization algorithm, to Dr. Axel Schleicher (C. and O. Vogt Brain Research Institute, University of Düsseldorf) for developing the observer-independent mapping technique, and to cand. med. Christian Grefkes for invaluable help with delineating the borders of area 6.

Over the years, the assistance of the technicians and undergraduate students of the C. and O. Vogt Brain Research Institute and Department of Neuroanatomy, University of Düsseldorf, has become indispensable to me. I would like to thank Ursula Blohm and Brigitte Machus for the technically demanding task of cutting entire human brains into sections 20 μm thick, Christine Opfermann-Rüngeler for excellent help with the artwork and Birgit Jansen with the photography of this book, and Sandra Lübke and Marion Mosch for technical assistance in realizing the project.

At the end of my acknowledgement list, one person is missing: Prof. Karl Zilles, my mentor in Düsseldorf. Since January 1993 it has been an honor for me to collaborate with a personality endowed with the quest for truth in science and research, integrity as a leader, and the commitment to continuously redefine one's own and others' limits. Now, as I am writing this in January 2003, I have spent a decade of my life in Düsseldorf—a decade that has been memorable and beautiful.

References

Aboitiz F, Garcia VR (1997) The evolutionary origin of the language areas in the human brain. A neuroanatomical perspective. Brain Res Rev 25:381–396

Alexander GE, Crutcher MD (1990) Preparation for movement: Neural representations of intended direction in three motor areas of the monkey. J Neurophysiol 64:133–150

Alexander GE, DeLong MR, Strick PL (1986) Parallel organization of functionally segregated circuits linking basal ganglia and cortex. Annu Rev Neurosci 9:357–381

Amunts K, Klingberg T, Binkofski F, Schormann T, Seitz RJ, Roland PE, Zilles K (1998) Cytoarchitectonic definition of Broca's region and its role in functions different from speech. Neuroimage 7:S8

Amunts K, Schleicher A, Bürgel U, Mohlberg H, Uylings HBM, Zilles K (1999) Broca's region revisited: Cytoarchitecture and intersubject variability. J Comp Neurol 412:319–341

Amunts K, Malikovic A, Mohlberg H, Schormann T, Zilles K (2000) Brodmann's areas 17 and 18 brought into stereotaxic space—Where and how variable? Neuroimage 11:66–84

Andersen RA, Asanuma C, Cowan WM (1985) Callosal and prefrontal associational projecting cell populations in area 7A of the macaque monkey: A study using retrogradely transported fluorescent dyes. J Comp Neurol 232:443–455

Andersen RA, Bracewell RM, Barash S, Gnadt JW, Fogassi L (1990) Eye position effects on visual, memory, and saccade-related activity in areas LIP and 7a of macaque. J Neurosci 10:1176–1196

Ashburner J, Friston K (1997) Multimodal image coregistration and partitioning—A unified framework. Neuroimage 6:209–217

Bailey P, von Bonin G (1951) The Isocortex of Man. University of Illinois Press, Urbana, Illinois

Barash S, Bracewell RM, Fogassi L, Gnadt JW, Andersen RA (1991) Saccade-related activity in the lateral intraparietal area II: Spatial properties. J Neurophysiol 66:1109–1124

Barbas H, Pandya DN (1987) Architecture and frontal cortical connections of the premotor cortex (area 6) in the rhesus monkey. J Comp Neurol 256:211–228

Bear MF, Connors BW, Paradiso MA (1996) Neuroscience: Exploring the Brain. Williams & Wilkins, Baltimore

Bernhard CG, Bohm E, Petersen I (1953) Investigations on the organization of the corticospinal system in monkeys (Macaca Mulatta). Acta Physiol Scand 29, Suppl. 106:79–105

Binkofski F, Buccino G, Posse S, Seitz RJ, Rizzolatti G, Freund HJ (1999) A fronto-parietal circuit for object manipulation in man: Evidence from an fMRI-study. Eur J Neurosci 11:3276–3286

Binkofski F, Amunts K, Stephan KM, Posse S, Schormann T, Freund HJ, Zilles K, Seitz RJ (2000a) Broca's region subserves imagery of motion: A combined cytoarchitectonic and fMRI study. Hum Brain Mapp 11:273–285

Binkofski F, Geyer S, Fink GR, Buccino G, Seitz RJ, Zilles K, Freund HJ (2000b) Differential activation of area 4p and 4a in motor-related attention. Neuroimage 11:S7

Binkofski F, Fink GR, Geyer S, Buccino G, Gruber O, Shah NJ, Taylor JG, Seitz RJ, Zilles K, Freund HJ (2002) Neural activity in human primary motor cortex areas 4a and 4p is modulated differentially by attention to action. J Neurophysiol 88:514–519

Blatt GJ, Andersen RA, Stoner GR (1990) Visual receptive field organization and cortico-cortical connections of the lateral intraparietal area (area LIP) in the macaque. J Comp Neurol 299:421–445

Bodegard A, Geyer S, Naito E, Zilles K, Roland PE (2000) Somatosensory areas in man activated by moving stimuli: Cytoarchitectonic mapping and PET. Neuroreport 11:187–191
Braak H (1980) Architectonics of the Human Telencephalic Cortex. Springer, Berlin
Bremmer F, Schlack A, Shah NJ, Zafiris O, Kubischik M, Hoffmann KP, Zilles K, Fink GR (2001) Polymodal motion processing in posterior parietal and premotor cortex: A human fMRI study strongly implies equivalencies between humans and monkeys. Neuron 29:287–296
Brinkman C (1981) Lesions in supplementary motor area interfere with a monkey's performance of a bimanual coordination task. Neurosci Lett 27:267–270
Brinkman C (1984) Supplementary motor area of the monkey's cerebral cortex: Short- and long-term deficits after unilateral ablation and the effects of subsequent callosal section. J Neurosci 4:918–929
Brinkman C, Porter R (1979) Supplementary motor area in the monkey: Activity of neurons during a learned motor task. J Neurophysiol 42:681–709
Brodmann K (1903) Beiträge zur histologischen Lokalisation der Großhirnrinde. Erste Mitteilung: Die Regio Rolandica. J Psychol Neurol 2:79–107
Brodmann K (1908) Beiträge zur histologischen Lokalisation der Großhirnrinde. Sechste Mitteilung: Die Cortexgliederung des Menschen. J Psychol Neurol 10:231–246
Brodmann K (1909) Vergleichende Lokalisationslehre der Großhirnrinde. Barth, Leipzig
Bruce CJ, Goldberg ME, Bushnell MC, Stanton GB (1985) Primate frontal eye fields. II. Physiological and anatomical correlates of electrically evoked eye movements. J Neurophysiol 54:714–734
Buccino G, Binkofski F, Fink GR, Fadiga L, Fogassi L, Gallese V, Seitz RJ, Zilles K, Rizzolatti G, Freund HJ (2001) Action observation activates premotor and parietal areas in a somatotopic manner: An fMRI study. Eur J Neurosci 13:400–404
Bucy PC (1934) The relation of the premotor cortex to motor activity. Journal of Nervous and Mental Disease 79:621–630
Bushnell MC, Goldberg ME, Robinson DL (1981) Behavioral enhancement of visual responses in monkey cerebral cortex. I: Modulation in posterior parietal cortex related to selective visual attention. J Neurophysiol 46:755–772
Caminiti R, Ferraina S, Johnson PB (1996) The sources of visual information to the primate frontal lobe: A novel role for the superior parietal lobule. Cereb Cortex 6:319–328
Campbell AW (1905) Histological Studies on the Localization of Cerebral Function. University Press, Cambridge
Chen LL, Wise SP (1995) Supplementary eye field contrasted with the frontal eye field during acquisition of conditional oculomotor associations. J Neurophysiol 73:1121–1133
Chen YC, Thaler D, Nixon PD, Stern CE, Passingham RE (1995) The functions of the medial premotor cortex II. The timing and selection of learned movements. Exp Brain Res 102:461–473
Cochin S, Barthelemy C, Roux S, Martineau J (1999) Observation and execution of movement: Similarities demonstrated by quantified electroencephalography. Eur J Neurosci 11:1839–1842
Colby CL, Duhamel JR (1991) Heterogeneity of extrastriate visual areas and multiple parietal areas in the macaque monkey. Neuropsychologia 29:517–537
Colby CL, Olson CR (1999) Spatial cognition. In: Zigmond MJ, Bloom FE, Landis SC, Roberts JL, Squire LR (eds) Fundamental Neuroscience. Academic Press, San Diego, pp 1363–1383
Colby CL, Gattass R, Olson CR, Gross CG (1988) Topographical organization of cortical afferents to extrastriate visual area PO in the macaque: A dual tracer study. J Comp Neurol 269:392–413
Colby CL, Duhamel JR, Goldberg ME (1993) Ventral intraparietal area of the macaque: Anatomic location and visual response properties. J Neurophysiol 69:902–914
Colby CL, Duhamel JR, Goldberg ME (1995) Oculocentric spatial representation in parietal cortex. Cereb Cortex 5:470–481
Colebatch JG, Deiber MP, Passingham RE, Friston KJ, Frackowiak RSJ (1991) Regional cerebral blood flow during voluntary arm and hand movements in human subjects. J Neurophysiol 65:1392–1401
Crivello F, Lemazurier N, Mazoyer N, Mazoyer B (1999) Comparing spatial normalization procedures. Neuroimage 9:S31

Crivello F, Schormann T, Tzourio-Mazoyer N, Roland PE, Zilles K, Mazoyer BM (2002) Comparison of spatial normalization procedures and their impact on functional maps. Hum Brain Mapp 16:228–250
Decety J, Perani D, Jeannerod M, Bettinardi V, Tadary B, Woods R, Mazziotta JC, Fazio F (1994) Mapping motor representations with positron emission tomography. Nature 371:600–602
Decety J, Grezes J, Costes N, Perani D, Jeannerod M, Procyk E, Grassi F, Fazio F (1997) Brain activity during observation of actions—Influence of action content and subject's strategy. Brain 120:1763–1777
Deiber MP, Passingham RE, Colebatch JG, Friston KJ, Nixon PD, Frackowiak RSJ (1991) Cortical areas and the selection of movement: A study with positron emission tomography. Exp Brain Res 84:393–402
Deiber MP, Wise SP, Honda M, Catalan MJ, Grafman J, Hallett M (1997) Frontal and parietal networks for conditional motor learning: A positron emission tomography study. J Neurophysiol 78:977–991
di Pellegrino G, Wise SP (1991) A neurophysiological comparison of three distinct regions of the primate frontal lobe. Brain 114:951–978
di Pellegrino G, Fadiga L, Fogassi L, Gallese V, Rizzolatti G (1992) Understanding motor events: A neurophysiological study. Exp Brain Res 91:176–180
Donoghue JP, Leibovic S, Sanes JN (1992) Organization of the forelimb area in squirrel monkey motor cortex: Representation of digit, wrist, and elbow muscles. Exp Brain Res 89:1–19
Duhamel JR, Colby CL, Goldberg ME (1991) Congruent representations of visual and somatosensory space in single neurons of monkey ventral intra-parietal cortex (area VIP). In: Paillard J (ed) Brain and Space. University Press, Oxford, pp 223–236
Duhamel JR, Colby CL, Goldberg ME (1992) The updating of the representation of visual space in parietal cortex by intended eye movements. Science 255:90–92
Dum RP, Strick PL (1991) The origin of corticospinal projections from the premotor areas in the frontal lobe. J Neurosci 11:667–689
Duvernoy HM, Delon S, Vannson JL (1981) Cortical blood vessels of the human brain. Brain Res Bull 7:519–579
Eccles JC (1982) The initiation of voluntary movements by the supplementary motor area. Arch Psychiatr Nervenkr 231:423–441
Ehrsson HH, Naito E, Geyer S, Amunts K, Zilles K, Forssberg H, Roland PE (2000) Simultaneous movements of upper and lower limbs are coordinated by motor representations that are shared by both limbs: A PET study. Eur J Neurosci 12:3385–3398
Evarts EV (1968) Relation of pyramidal tract activity to force exerted during voluntary movement. J Neurophysiol 31:14–27
Fadiga L, Fogassi L, Pavesi G, Rizzolatti G (1995) Motor facilitation during action observation: A magnetic stimulation study. J Neurophysiol 73:2608–2611
Ferraina S, Garasto MR, Battaglia-Mayer A, Ferraresi P, Johnson PB, Lacquaniti F, Caminiti R (1997a) Visual control of hand-reaching movement: Activity in parietal area 7m. Eur J Neurosci 9:1090–1095
Ferraina S, Johnson PB, Garasto MR, Battaglia-Mayer A, Ercolani L, Bianchi L, Lacquaniti F, Caminiti R (1997b) Combination of hand and gaze signals during reaching: Activity in parietal area 7m of the monkey. J Neurophysiol 77:1034–1038
Ferrier D (1876) The Functions of the Brain. Smith, Elder and Company, London
Ferrier D (1886) The Functions of the Brain. G. P. Putnam's Sons, New York
Ferrier D, Yeo G (1885) A record of experiments on the effects of lesions of different regions of the cerebral hemispheres. Philosophical Transactions of the Royal Society of London (Biology) 175:479–564
Finger S (1994) Origins of Neuroscience—A History of Explorations into Brain Function. Oxford University Press, New York
Fink GR, Frackowiak RSJ, Pietrzyk U, Passingham RE (1997) Multiple nonprimary motor areas in the human cortex. J Neurophysiol 77:2164–2174
Foerster O (1931) The cerebral cortex in man. Lancet 2:309–312
Foerster O (1936) Motorische Felder und Bahnen. In: Bumke O, Foerster O (eds) Handbuch der Neurologie. Springer, Berlin, pp 1–357

Fogassi L, Gallese V, Fadiga L, Luppino G, Matelli M, Rizzolatti G (1996) Coding of peripersonal space in inferior premotor cortex (area F4). J Neurophysiol 76:141–157

Fogassi L, Gallese V, Fadiga L, Rizzolatti G (1998) Neurons responding to the sight of goal-directed hand/arm actions in the parietal area PF (7b) of the macaque monkey. Soc Neurosci Abstr 24/1:654

Fogassi L, Raos V, Franchi G, Gallese V, Luppino G, Matelli M (1999) Visual responses in the dorsal premotor area F2 of the macaque monkey. Exp Brain Res 128:194–199

Franz SI, Lashley KS (1917) The retention of habits by the rat after destruction of the frontal portion of the cerebrum. Psychobiology 1:3–18

Freund HJ (1996) Historical overview. In: Lüders HO (ed) Supplementary Sensorimotor Area. Lippincott-Raven, Philadelphia, pp 17–27

Freund HJ, Hummelsheim H (1985) Lesions of premotor cortex in man. Brain 108:697–733

Fried I (1996) Electrical stimulation of the supplementary sensorimotor area. In: Lüders HO (ed) Supplementary Sensorimotor Area. Lippincott-Raven, Philadelphia, pp 177–185

Fries W (1985) Inputs from motor and premotor cortex to the superior colliculus of the macaque monkey. Behav Brain Res 18:95–105

Fritsch G, Hitzig E (1870) Über die elektrische Erregbarkeit des Großhirns. Archiv für Anatomie und Physiologie 37:300–332

Fulton JF (1937) Spasticity and the frontal lobes. A review. N Engl J Med 217:1017–1024

Gabernet L, Meskenaite V, Hepp-Reymond MC (1999) Parcellation of the lateral premotor cortex of the macaque monkey based on staining with the neurofilament antibody SMI-32. Exp Brain Res 128:188–193

Gallese V, Murata A, Kaseda M, Niki N, Sakata H (1994) Deficit of hand preshaping after muscimol injection in monkey parietal cortex. Neuroreport 5:1525–1529

Gallese V, Fadiga L, Fogassi L, Rizzolatti G (1996) Action recognition in the premotor cortex. Brain 119:593–609

Gallese V, Fadiga L, Fogassi L, Luppino G, Murata A (1997) A parietofrontal circuit for hand grasping movements in the monkey: Evidence from reversible inactivation experiments. In: Thier P, Karnath HO (eds) Parietal Lobe Contributions to Orientation in 3D Space. Springer, Heidelberg, pp 255–270

Galletti C, Fattori P, Battaglini PP, Shipp S, Zeki S (1996) Functional demarcation of a border between areas V6 and V6A in the superior parietal gyrus of the macaque monkey. Eur J Neurosci 8:30–52

Galletti C, Fattori P, Kutz DF, Battaglini PP (1997) Arm movement-related neurons in the visual area V6A of the macaque superior parietal lobule. Eur J Neurosci 9:410–413

Garey LJ (1994) Brodmann's "Localisation in the Cerebral Cortex". Smith-Gordon, London

Gentilucci M, Fogassi L, Luppino G, Matelli M, Camarda R, Rizzolatti G (1988) Functional organization of inferior area 6 in the macaque monkey. I. Somatotopy and the control of proximal movements. Exp Brain Res 71:475–490

Gentilucci M, Fogassi L, Luppino G, Matelli M, Camarda R, Rizzolatti G (1989) Somatotopic representation in inferior area 6 of the macaque monkey. Brain Behav Evol 33:118–121

Georgopoulos AP, Kalaska JF, Caminiti R, Massey JT (1982) On the relations between the direction of two-dimensional arm movements and cell discharge in primate motor cortex. J Neurosci 2:1527–1537

Gerardin E, Sirigu A, Lehericy S, Poline JB, Gaymard B, Marsault C, Agid Y, Le Bihan D (2000) Partially overlapping neural networks for real and imagined hand movements. Cereb Cortex 10:1093–1104

Geyer S, Ledberg A, Schleicher A, Kinomura S, Schormann T, Bürgel U, Klingberg T, Larsson J, Zilles K, Roland PE (1996) Two different areas within the primary motor cortex of man. Nature 382:805–807

Geyer S, Schleicher A, Zilles K (1997) The somatosensory cortex of human: Cytoarchitecture and regional distributions of receptor-binding sites. Neuroimage 6:27–45

Geyer S, Matelli M, Luppino G, Schleicher A, Jansen Y, Palomero-Gallagher N, Zilles K (1998a) Receptor autoradiographic mapping of the mesial motor and premotor cortex of the macaque monkey. J Comp Neurol 397:231–250

Geyer S, Matelli M, Luppino G, Zilles K (1998b) A new microstructural map of the macaque monkey lateral premotor cortex based on neurofilament protein distribution. Eur J Neurosci 10, Suppl.10:83

Geyer S, Schleicher A, Zilles K (1999) Areas 3a, 3b, and 1 of human primary somatosensory cortex: 1. Microstructural organization and interindividual variability. Neuroimage 10:63–83

Geyer S, Matelli M, Luppino G, Zilles K (2000a) Functional neuroanatomy of the primate isocortical motor system. Anat Embryol 202:443–474

Geyer S, Schormann T, Mohlberg H, Zilles K (2000b) Areas 3a, 3b, and 1 of human primary somatosensory cortex: 2. Spatial normalization to standard anatomical space. Neuroimage 11:684–696

Geyer S, Zilles K, Luppino G, Matelli M (2000c) Neurofilament protein distribution in the macaque monkey dorsolateral premotor cortex. Eur J Neurosci 12:1554–1566

Ghosh S, Brinkman C, Porter R (1987) A quantitative study of the distribution of neurons projecting to the precentral motor cortex in the monkey (M. fascicularis). J Comp Neurol 259:424–444

Gnadt JW, Andersen RA (1988) Memory related motor planning activity in posterior parietal cortex of macaque. Exp Brain Res 70:216–220

Godschalk M, Mitz AR, van Duin B, van der Burg H (1995) Somatotopy of monkey premotor cortex examined with microstimulation. Neurosci Res 23:269–279

Goldberg G (1985) Supplementary motor area structure and function: Review and hypotheses. Behav Brain Sci 8:567–616

Goldberg ME, Colby CL, Duhamel JR (1990) Representation of visuomotor space in the parietal lobe of the monkey. Cold Spring Harbor Symp 55:729–739

Goldman-Rakic PS, Selemon LD (1986) Topography of projections in non human primates and implications for functional parcellation of the neostriatum. In: Peters A, Jones EG (eds) Cerebral Cortex, Vol. 5. Plenum, New York, pp 447–466

Gould HJ, Cusick CG, Pons TP, Kaas JH (1986) The relationship of corpus callosum connections to electrical stimulation maps of motor, supplementary motor, and the frontal eye fields in owl monkeys. J Comp Neurol 247:297–325

Grafton ST, Mazziotta JC, Presty S, Friston KJ, Frackowiak RSJ, Phelps ME (1992a) Functional anatomy of human procedural learning determined with regional cerebral blood flow and PET. J Neurosci 12:2542–2548

Grafton ST, Mazziotta JC, Woods RP, Phelps ME (1992b) Human functional anatomy of visually guided finger movements. Brain 115:565–587

Grafton ST, Arbib MA, Fadiga L, Rizzolatti G (1996a) Localization of grasp representations in humans by positron emission tomography. 2. Observation compared with imagination. Exp Brain Res 112:103–111

Grafton ST, Fagg AH, Woods RP, Arbib MA (1996b) Functional anatomy of pointing and grasping in humans. Cereb Cortex 6:226–237

Grafton ST, Fagg AH, Arbib MA (1998) Dorsal premotor cortex and conditional movement selection: A PET functional mapping study. J Neurophysiol 79:1092–1097

Graziano MSA, Yap GS, Gross CG (1994) Coding of visual space by premotor neurons. Science 266:1054–1057

Grezes J, Costes N, Decety J (1998) Top down effect of strategy on the perception of human biological motion: A PET investigation. Cognitive Neuropsychol 15:553–582

Grosbras MH, Lobel E, van de Moortele PF, Le Bihan D, Berthoz A (1999) An anatomical landmark for the supplementary eye fields in human revealed with functional magnetic resonance imaging. Cereb Cortex 9:705–711

Grünbaum ASF, Sherrington CS (1902) Observations on the physiology of the cerebral cortex in some of the higher apes. Proceedings of the Royal Society 69:206–209

Grünbaum ASF, Sherrington CS (1903) Observations on the physiology of the cerebral cortex in anthropoid apes. Proceedings of the Royal Society 72:152–155

Hallett M, Fieldman J, Cohen LG, Sadato N, Pascual-Leone A (1994) Involvement of primary motor cortex in motor imagery and mental practice. Behav Brain Sci 17:210

Halsband U (1982) D. Phil. Thesis: Higher Movement Disorders in Monkeys. University of Oxford,

Halsband U, Freund HJ (1990) Premotor cortex and conditional motor learning in man. Brain 113:207–222
Halsband U, Passingham RE (1982) The role of premotor and parietal cortex in the direction of action. Brain Res 240:368–372
Hanes DP, Thompson KG, Schall JD (1995) Relationship of presaccadic activity in frontal eye field and supplementary eye field to saccade initiation in macaque: Poisson spike train analysis. Exp Brain Res 103:85–96
Hari R, Forss N, Avikainen S, Kirveskari E, Salenius S, Rizzolatti G (1998) Activation of human primary motor cortex during action observation: A neuromagnetic study. Proc Natl Acad Sci USA 95:15061–15065
He SQ, Dum RP, Strick PL (1993) Topographic organization of corticospinal projections from the frontal lobe: Motor areas on the lateral surface of the hemisphere. J Neurosci 13:952–980
He SQ, Dum RP, Strick PL (1995) Topographic organization of corticospinal projections from the frontal lobe: Motor areas on the medial surface of the hemisphere. J Neurosci 15:3284–3306
Hepp-Reymond MC, Hüsler EJ, Maier MA, Qi HX (1994) Force-related neuronal activity in two regions of the primate ventral premotor cortex. Can J Physiol Pharmacol 72:571–579
Hikosaka O, Sakai K, Nakahara H, Lu X, Miyachi S, Nakamura K, Rand MK (1999) Neural mechanisms for learning of sequential procedures. In: Gazzaniga MS (ed) The New Cognitive Neurosciences. MIT Press, Cambridge, pp 553–572
Hitzig E (1904) Untersuchungen über das Gehirn. A. Hirschwald, Berlin
Hof PR, Nimchinsky EA, Morrison JH (1995) Neurochemical phenotype of corticocortical connections in the macaque monkey: Quantitative analysis of a subset of neurofilament protein-immunoreactive projection neurons in frontal, parietal, temporal, and cingulate cortices. J Comp Neurol 362:109–133
Holmes G, May WP (1909) On the exact origin of the pyramidal tracts in man and other mammals. Brain 32:1–43
Huerta MF, Kaas JH (1990) Supplementary eye field as defined by intracortical microstimulation: Connections in macaques. J Comp Neurol 293:299–330
Hummelsheim H, Bianchetti M, Wiesendanger M, Wiesendanger R (1988) Sensory input to the agranular motor fields: A comparison between precentral, supplementary-motor and premotor areas in the monkey. Exp Brain Res 69:289–298
Iacoboni M, Woods RP, Mazziotta JC (1998) Bimodal (auditory and visual) left frontoparietal circuitry for sensorimotor integration and sensorimotor learning. Brain 121:2135–2143
Iacoboni M, Woods RP, Brass M, Bekkering H, Mazziotta JC, Rizzolatti G (1999) Cortical mechanisms of human imitation. Science 286:2526–2528
Indefrey P, Brown CM, Hagoort P, Amunts K, Hellwig F, Herzog H, Seitz RJ (1999) Noun phrase and sentence-level syntactic encoding during language production activate the left frontal operculum—A PET study. Neuroimage 9:S1027
Iwamura Y, Tanaka M (1996) Representation of reaching and grasping in the monkey postcentral gyrus. Neurosci Lett 214:147–150
Jackson JH (1863) Convulsive spasms of the right hand and arm preceding epileptic seizures. Medical Times and Gazette 1:110–111
Jeannerod M (1994) The representing brain: Neural correlates of motor intention and imagery. Behav Brain Sci 17:187–245
Jeannerod M, Decety J (1995) Mental motor imagery: A window into the representational stages of action. Curr Opin Neurobiol 5:727–732
Jeannerod M, Arbib MA, Rizzolatti G, Sakata H (1995) Grasping objects: The cortical mechanisms of visuomotor transformation. Trends Neurosci 18:314–320
Jenkins IH, Brooks DJ, Nixon PD, Frackowiak RSJ, Passingham RE (1994) Motor sequence learning: A study with positron emission tomography. J Neurosci 14:3775–3790
Johnson PB, Ferraina S, Bianchi L, Caminiti R (1996) Cortical networks for visual reaching: Physiological and anatomical organization of frontal and parietal lobe arm regions. Cereb Cortex 6:102–119
Jones EG, Powell TPS (1969) Connexions of the somatic sensory cortex of the rhesus monkey. I. Ipsilateral cortical connexions. Brain 92:477–502

Jones EG, Wise SP (1977) Size, laminar and columnar distribution of efferent cells in the sensory-motor cortex of monkeys. J Comp Neurol 175:391–437

Kalaska JF, Cohen DA, Prud'homme M, Hyde ML (1990) Parietal area 5 neuronal activity encodes movement kinematics, not movement dynamics. Exp Brain Res 80:351–364

Kawashima R, Roland PE, O'Sullivan BT (1994) Fields in human motor areas involved in preparation for reaching, actual reaching, and visuomotor learning: A positron emission tomography study. J Neurosci 14:3462–3474

Kawashima R, Roland PE, O'Sullivan BT (1995) Functional anatomy of reaching and visuomotor learning: A positron emission tomography study. Cereb Cortex 5:111–122

Keizer K, Kuypers HGJM (1989) Distribution of corticospinal neurons with collaterals to the lower brain stem reticular formation in monkey (Macaca fascicularis). Exp Brain Res 74:311–318

Krams M, Rushworth MFS, Deiber MP, Frackowiak RSJ, Passingham RE (1998) The preparation, execution and suppression of copied movements in the human brain. Exp Brain Res 120:386–398

Kurata K (1989) Distribution of neurons with set- and movement-related activity before hand and foot movements in the premotor cortex of rhesus monkey. Exp Brain Res 77:245–256

Kurata K, Tanji J (1986) Premotor cortex neurons in macaques: Activity before distal and proximal forelimb movements. J Neurosci 6:403–411

Kurata K, Wise SP (1988) Premotor and supplementary motor cortex in rhesus monkeys: Neuronal activity during externally and internally instructed motor tasks. Exp Brain Res 72:237–248

Kuypers HGJM (1962) Corticospinal connections: Postnatal development in the rhesus monkey. Science 138:678–680

Kuypers HGJM (1981) Anatomy of the descending pathways. In: Brookhart JM, Mountcastle VB (eds) Handbook of Physiology—The Nervous System II. American Physiological Society, Bethesda, pp 597–666

Lashley KS, Clark G (1946) The cytoarchitecture of the cerebral cortex of Ateles: A critical examination of architectonic studies. J Comp Neurol 85:223–305

Lashley KS, Franz SI (1917) The effects of cerebral destruction upon habit-formation and retention in the albino rat. Psychobiology 1:71–139

Lee VMY, Otvos L, Carden MJ, Hollosi M, Dietzschold B, Lazzarini RA (1988) Identification of the major multiphosphorylation site in mammalian neurofilaments. Proc Natl Acad Sci USA 85:1998–2002

Leonardo M, Fieldman J, Sadato N, Campbell G, Ibanez V, Cohen L, Deiber MP, Jezzard P, Pons T, Turner R, Le Bihan D, Hallett M (1995) A functional magnetic resonance imaging study of cortical regions associated with motor task execution and motor ideation in humans. Hum Brain Mapp 3:83–92

Lim SH, Dinner DS, Lüders HO (1996) Cortical stimulation of the supplementary sensorimotor area. In: Lüders HO (ed) Supplementary Sensorimotor Area. Lippincott-Raven, Philadelphia, pp 187–197

Luppino G, Matelli M, Rizzolatti G (1990) Cortico-cortical connections of two electrophysiologically identified arm representations in the mesial agranular frontal cortex. Exp Brain Res 82:214–218

Luppino G, Matelli M, Camarda RM, Gallese V, Rizzolatti G (1991) Multiple representations of body movements in mesial area 6 and the adjacent cingulate cortex: An intracortical microstimulation study in the macaque monkey. J Comp Neurol 311:463–482

Luppino G, Matelli M, Camarda R, Rizzolatti G (1993) Corticocortical connections of area F3 (SMA-proper) and area F6 (pre-SMA) in the macaque monkey. J Comp Neurol 338:114–140

Luppino G, Matelli M, Camarda R, Rizzolatti G (1994) Corticospinal projections from mesial frontal and cingulate areas in the monkey. Neuroreport 5:2545–2548

Luppino G, Murata A, Govoni P, Matelli M (1999) Largely segregated parietofrontal connections linking rostral intraparietal cortex (areas AIP and VIP) and the ventral premotor cortex (areas F5 and F4). Exp Brain Res 128:181–187

Mahalanobis PC, Majumda DN, Rao CR (1949) Anthropometric survey of the united provinces. A statistical study. Sankhya 9:89–324

Matelli M, Luppino G (1996) Thalamic input to mesial and superior area 6 in the macaque monkey. J Comp Neurol 372:59–87
Matelli M, Luppino G, Rizzolatti G (1985) Patterns of cytochrome oxidase activity in the frontal agranular cortex of the macaque monkey. Behav Brain Res 18:125–136
Matelli M, Camarda R, Glickstein M, Rizzolatti G (1986) Afferent and efferent projections of the inferior area 6 in the macaque monkey. J Comp Neurol 251:281–298
Matelli M, Luppino G, Fogassi L, Rizzolatti G (1989) Thalamic input to inferior area 6 and area 4 in the macaque monkey. J Comp Neurol 280:468–488
Matelli M, Luppino G, Rizzolatti G (1991) Architecture of superior and mesial area 6 and the adjacent cingulate cortex in the macaque monkey. J Comp Neurol 311:445–462
Matelli M, Luppino G, Govoni P, Geyer S (1996) Anatomical and functional subdivisions of inferior area 6 in macaque monkey. Soc Neurosci Abstr 22/3:2024
Matelli M, Govoni P, Galletti C, Kutz DF, Luppino G (1998) Superior area 6 afferents from the superior parietal lobule in the macaque monkey. J Comp Neurol 402:327–352
Matsuzaka Y, Tanji J (1996) Changing directions of forthcoming arm movements: Neuronal activity in the presupplementary and supplementary motor area of monkey cerebral cortex. J Neurophysiol 76:2327–2342
Matsuzaka Y, Aizawa H, Tanji J (1992) A motor area rostral to the supplementary motor area (presupplementary motor area) in the monkey: Neuronal activity during a learned motor task. J Neurophysiol 68:653–662
Merker B (1983) Silver staining of cell bodies by means of physical development. J Neurosci Meth 9:235–241
Mitz AR, Wise SP (1987) The somatotopic organization of the supplementary motor area: Intracortical microstimulation mapping. J Neurosci 7:1010–1021
Mountcastle VB, Lynch JC, Georgopoulos A, Sakata H, Acuna C (1975) Posterior parietal association cortex of the monkey: Command functions for operations within extrapersonal space. J Neurophysiol 38:871–908
Murata A, Gallese V, Kaseda M, Sakata H (1996) Parietal neurons related to memory-guided hand manipulation. J Neurophysiol 75:2180–2186
Murata A, Fadiga L, Fogassi L, Gallese V, Raos V, Rizzolatti G (1997) Object representation in the ventral premotor cortex (area F5) of the monkey. J Neurophysiol 78:2226–2230
Murray EA, Coulter JD (1981) Organization of corticospinal neurons in the monkey. J Comp Neurol 195:339–365
Mushiake H, Inase M, Tanji J (1991) Neuronal activity in the primate premotor, supplementary, and precentral motor cortex during visually guided and internally determined sequential movements. J Neurophysiol 66:705–718
Naito E, Ehrsson HH, Geyer S, Zilles K, Roland PE (1999) Illusory arm movements activate cortical motor areas: A positron emission tomography study. J Neurosci 19:6134–6144
Naito E, Kinomura S, Geyer S, Kawashima R, Roland PE, Zilles K (2000) Fast reaction to different sensory modalities activates common fields in the motor areas, but the anterior cingulate cortex is involved in the speed of reaction. J Neurophysiol 83:1701-1709
Okano K, Tanji J (1987) Neuronal activities in the primate motor fields of the agranular frontal cortex preceding visually triggered and self-paced movement. Exp Brain Res 66:155–166
Olson CR, Gettner SN (1995) Object-centered direction selectivity in the macaque supplementary eye field. Science 269:985–988
Olson CR, Gettner SN (1996) Brain representation of object-centered space. Curr Opin Neurobiol 6:165–170
Olszewski J (1952) The Thalamus of *Macaca mulatta*. Karger, New York
Ono M, Kubik S, Abernathey CD (1990) Atlas of the Cerebral Sulci. Thieme, Stuttgart
Pandya DN, Seltzer B (1982) Intrinsic connections and architectonics of posterior parietal cortex in the rhesus monkey. J Comp Neurol 204:196–210
Parsons LM, Fox PT, Downs JH, Glass T, Hirsch TB, Martin CC, Jerabek PA, Lancaster JL (1995) Use of implicit motor imagery for visual shape discrimination as revealed by PET. Nature 375:54–58
Passingham RE (1993) The Frontal Lobes and Voluntary Action. University Press, Oxford

Passingham RE, Chen YC, Thaler D (1989) Supplementary motor cortex and self-initiated movement. In: Ito M (ed) Neural Programming. Karger, Tokyo, pp 13–24
Paus T (1996) Location and function of the human frontal eye-field: A selective review. Neuropsychologia 34:475–483
Penfield W, Boldrey E (1937) Somatic motor and sensory representation in the cerebral cortex of man as studied by electrical stimulation. Brain 60:389–443
Penfield W, Rasmussen T (1952) The Cerebral Cortex of Man. Macmillan, New York
Penfield W, Welch K (1951) The supplementary motor area of the cerebral cortex: A clinical and experimental study. Arch Neurol Psychiat 66:289–317
Petit L, Courtney SM, Ungerleider LG, Haxby JV (1998) Sustained activity in the medial wall during working memory delays. J Neurosci 18:9429–9437
Petrides M (1982) Motor conditional associative-learning after selective prefrontal lesions in the monkey. Behav Brain Res 5:407–413
Petrides M, Paxinos G, Huang XF, Morris R, Pandya DN (2000) Delineation of the monkey cortex on the basis of the distribution of a neurofilament protein. In: Paxinos G, Huang XF, Toga AW (eds) The Rhesus Monkey Brain in Stereotaxic Coordinates. Academic Press, San Diego, pp 155–229
Picard N, Strick PL (1996) Motor areas of the medial wall: A review of their location and functional activation. Cereb Cortex 6:342–353
Pierrot-Deseilligny C, Rivaud S, Gaymard B, Müri R, Vermersch AI (1995) Cortical control of saccades. Ann Neurol 37:557–567
Poliakov AV, Schieber MH (1999) Limited functional grouping of neurons in the motor cortex hand area during individuated finger movements: A cluster analysis. J Neurophysiol 82:3488–3505
Porro CA, Francescato MP, Cettolo V, Diamond ME, Baraldi P, Zuiani C, Bazzocchi M, Di Prampero PE (1996) Primary motor and sensory cortex activation during motor performance and motor imagery: A functional magnetic resonance imaging study. J Neurosci 16:7688–7698
Porter R, Lemon R (1993) Corticospinal Function and Voluntary Movement. Oxford University Press, Oxford
Rademacher J, Caviness VS, Steinmetz H, Galaburda AM (1993) Topographical variation of the human primary cortices: Implications for neuroimaging, brain mapping, and neurobiology. Cereb Cortex 3:313–329
Rajkowska G, Goldman-Rakic PS (1995) Cytoarchitectonic definition of prefrontal areas in the normal human cortex: II. Variability in locations of areas 9 and 46 and relationship to the Talairach coordinate system. Cereb Cortex 5:323–337
Rao SM, Binder JR, Bandettini PA, Hammeke TA, Yetkin FZ, Jesmanowicz A, Lisk LM, Morris GL, Mueller WM, Estkowski LD, Wong EC, Haughton VM, Hyde JS (1993) Functional magnetic resonance imaging of complex human movements. Neurology 43:2311–2318
Rizzolatti G, Matelli M, Pavesi G (1983) Deficits in attention and movement following the removal of postarcuate (area 6) and prearcuate (area 8) cortex in macaque monkeys. Brain 106:655–673
Rizzolatti G, Camarda R, Fogassi L, Gentilucci M, Luppino G, Matelli M (1988) Functional organization of inferior area 6 in the macaque monkey. II. Area F5 and the control of distal movements. Exp Brain Res 71:491–507
Rizzolatti G, Gentilucci M, Camarda RM, Gallese V, Luppino G, Matelli M, Fogassi L (1990) Neurons related to reaching-grasping arm movements in the rostral part of area 6 (area 6aβ). Exp Brain Res 82:337–350
Rizzolatti G, Fadiga L, Gallese V, Fogassi L (1996a) Premotor cortex and the recognition of motor actions. Cognitive Brain Res 3:131–141
Rizzolatti G, Fadiga L, Matelli M, Bettinardi V, Paulesu E, Perani D, Fazio F (1996b) Localization of grasp representations in humans by PET: 1. Observation versus execution. Exp Brain Res 111:246–252
Rizzolatti G, Luppino G, Matelli M (1996c) The classic supplementary motor area is formed by two independent areas. In: Lüders HO (ed) Supplementary Sensorimotor Area. Lippincott-Raven, Philadelphia, pp 45–56

Rizzolatti G, Fogassi L, Gallese V (1997) Parietal cortex: From sight to action. Curr Opin Neurobiol 7:562–567
Rizzolatti G, Luppino G, Matelli M (1998) The organization of the cortical motor system: New concepts. Electroencephalogr Clin Neurophysiol 106:283–296
Robinson DL, Goldberg ME, Stanton GB (1978) Parietal association cortex in the primate: Sensory mechanisms and behavioral modulations. J Neurophysiol 41:910–932
Roland PE, Zilles K (1994) Brain atlases—a new research tool. Trends Neurosci 17:458–467
Roland PE, Zilles K (1996) Functions and structures of the motor cortices in humans. Curr Opin Neurobiol 6:773–781
Roland PE, Larsen B, Lassen NA, Skinhoj E (1980) Supplementary motor area and other cortical areas in organization of voluntary movements in man. J Neurophysiol 43:118–136
Roland PE, Graufelds CJ, Wåhlin J, Ingelman L, Andersson M, Ledberg A, Pedersen J, Åkerman S, Dabringhaus A, Zilles K (1994) Human brain atlas: For high-resolution functional and anatomical mapping. Hum Brain Mapp 1:173–184
Roland PE, Geyer S, Amunts K, Schormann T, Schleicher A, Malikovic A, Zilles K (1997) Cytoarchitectural maps of the human brain in standard anatomical space. Hum Brain Mapp 5:222–227
Romo R, Schultz W (1987) Neuronal activity preceding self-initiated or externally timed arm movements in area 6 of monkey cortex. Exp Brain Res 67:656–662
Roth M, Decety J, Raybaudi M, Massarelli R, Delon-Martin C, Segebarth C, Morand S, Gemignani A, Décorps M, Jeannerod M (1996) Possible involvement of primary motor cortex in mentally simulated movement: A functional magnetic resonance imaging study. Neuroreport 7:1280–1284
Rouiller EM, Liang F, Babalian A, Moret V, Wiesendanger M (1994) Cerebellothalamocortical and pallidothalamocortical projections to the primary and supplementary motor cortical areas: A multiple tracing study in macaque monkeys. J Comp Neurol 345:185–213
Russel JR, DeMeyer W (1961) The quantitative cortical origin of pyramidal axons of Macaca rhesus, with some remarks on the slow rate of axolysis. Neurology 11:96–108
Sabbah P, Simond G, Levrier O, Habib M, Trabaud V, Murayama N, Mazoyer BM, Briant JF, Raybaud C, Salamon G (1995) Functional magnetic resonance imaging at 1.5 T during sensorimotor and cognitive task. Eur Neurol 35:131–136
Sakata H (1996) Coding of 3-D features of objects used in manipulation by parietal neurons. In: Caminiti R, Hoffmann KP, Lacquaniti F, Altman J (eds) Vision and Movement Mechanisms in the Cerebral Cortex. Human Frontier Science Program, Strasbourg, pp 55–63
Sakata H, Taira M, Murata A, Mine S (1995) Neural mechanisms of visual guidance of hand action in the parietal cortex of the monkey. Cereb Cortex 5:429–438
Sanes JN (1994) Neurophysiology of preparation, movement and imagery. Behav Brain Sci 17:221–223
Sanes JN, Donoghue JP, Thangaraj V, Edelman RR, Warach S (1995) Shared neural substrates controlling hand movements in human motor cortex. Science 268:1775–1777
Sarkissov SA, Filimonoff IN, Kononowa EP, Preobraschenskaja IS, Kukuew LA (1955) Atlas of the Cytoarchitectonics of the Human Cerebral Cortex. Medgiz, Moscow
Schieber MH (1999) Voluntary descending control. In: Zigmond MJ, Bloom FE, Landis SC, Roberts JL, Squire LR (eds) Fundamental Neuroscience. Academic Press, San Diego, pp 931–949
Schieber MH, Hibbard LS (1993) How somatotopic is the motor cortex hand area? Science 261:489–492
Schlag J, Schlag-Rey M (1987) Evidence for a supplementary eye field. J Neurophysiol 57:179–200
Schlaug G, Knorr U, Seitz RJ (1994) Inter-subject variability of cerebral activations in acquiring a motor skill: A study with positron emission tomography. Exp Brain Res 98:523–534
Schleicher A, Zilles K (1990) A quantitative approach to cytoarchitectonics: Analysis of structural inhomogeneities in nervous tissue using an image analyser. J Microsc 157:367–381
Schleicher A, Amunts K, Geyer S, Morosan P, Zilles K (1999) Observer-independent method for microstructural parcellation of cerebral cortex: A quantitative approach to cytoarchitectonics. Neuroimage 9:165–177

Schormann T, Zilles K (1997) Limitations of the principal-axes theory. IEEE Trans Med Imaging 16:942–947

Schormann T, Zilles K (1998) Three-dimensional linear and nonlinear transformations: An integration of light microscopical and MRI data. Hum Brain Mapp 6:339–347

Schormann T, von Matthey M, Dabringhaus A, Zilles K (1993) Alignment of 3-D brain data sets originating from MR and histology. Bioimaging 1:119–128

Schormann T, Dabringhaus A, Zilles K (1995) Statistics of deformations in histology and application to improved alignment with MRI. IEEE Trans Med Imaging 14:25–35

Schormann T, Henn S, Zilles K (1996) A new approach to fast elastic alignment with application to human brains. Lect Notes Comput Sci 1131:437–442

Schormann T, Dabringhaus A, Zilles K (1997) Extension of the principal-axes theory for the determination of affine transformations. In: Paulus E, Wahl FM (eds) Mustererkennung 1997 (19. DAGM-Symposium, Braunschweig, 15.-17. September 1997). Springer, Berlin, pp 384–391

Seitz RJ, Roland PE (1992) Learning of sequential finger movements in man: A combined kinematic and positron emission tomography (PET) study. Eur J Neurosci 4:154–165

Seitz RJ, Canavan AGM, Yágüez L, Herzog H, Tellmann L, Knorr U, Huang Y, Hömberg V (1997) Representations of graphomotor trajectories in the human parietal cortex: Evidence for controlled processing and automatic performance. Eur J Neurosci 9:378–389

Sergent J, Zuck E, Terriah S, MacDonald B (1992) Distributed neural network underlying musical sight-reading and keyboard performance. Science 257:106–109

Shima K, Mushiake H, Saito N, Tanji J (1996) Role for cells in the presupplementary motor area in updating motor plans. Proc Natl Acad Sci USA 93:8694–8698

Shinoda Y, Yokota JI, Futami T (1981) Divergent projection of individual corticospinal axons to motoneurons of multiple muscles in the monkey. Neurosci Lett 23:7–12

Sloper JJ, Powell TPS (1979) An experimental electron microscopic study of afferent connections to the primate motor and somatic sensory cortices. Phil Trans Roy Soc London B 285:199–226

Smith GE (1907) A new topographical survey of the human cerebral cortex, being an account of the distribution of the anatomically distinct cortical areas and their relationship to the cerebral sulci. J Anat 41:237–254

Stephan KM, Fink GR, Passingham RE, Silbersweig D, Ceballos-Baumann AO, Frith CD, Frackowiak RSJ (1995) Functional anatomy of the mental representation of upper extremity movements in healthy subjects. J Neurophysiol 73:373–386

Stepniewska I, Preuss TM, Kaas JH (1993) Architectonics, somatotopic organization, and ipsilateral cortical connections of the primary motor area (M1) of owl monkeys. J Comp Neurol 330:238–271

Sternberger LA, Sternberger NH (1983) Monoclonal antibodies distinguish phosphorylated and nonphosphorylated forms of neurofilaments *in situ*. Proc Natl Acad Sci USA 80:6126–6130

Strafella AP, Paus T (2000) Modulation of cortical excitability during action observation: A transcranial magnetic stimulation study. Neuroreport 11:2289–2292

Strick PL, Sterling P (1974) Synaptic termination of afferents from the ventrolateral nucleus of the thalamus in the cat motor cortex. A light and electron microscope study. J Comp Neurol 153:77–106

Suzuki H, Azuma M (1983) Topographic studies on visual neurons in the dorsolateral prefrontal cortex of the monkey. Exp Brain Res 53:47–58

Taira M, Mine S, Georgopoulos AP, Murata A, Sakata H (1990) Parietal cortex neurons of the monkey related to the visual guidance of hand movement. Exp Brain Res 83:29–36

Talairach J, Tournoux P (1988) Co-Planar Stereotaxic Atlas of the Human Brain. 3-Dimensional Proportional System: An Approach to Cerebral Imaging. Thieme, Stuttgart

Tanji J (1994) The supplementary motor area in the cerebral cortex. Neurosci Res 19:251–268

Tanji J (1996) New concepts of the supplementary motor area. Curr Opin Neurobiol 6:782–787

Tanji J, Kurata K (1979) Neuronal activity in the cortical supplementary motor area related with distal and proximal forelimb movements. Neurosci Lett 12:201–206

Tanji J, Kurata K (1982) Comparison of movement-related activity in two cortical areas of primates. J Neurophysiol 48:633–653

Tanji J, Kurata K (1985) Contrasting neuronal activity in supplementary and precentral motor cortex of monkeys. I. Responses to instructions determining motor responses to forthcoming signals of different modalities. J Neurophysiol 53:129–141

Tanji J, Shima K (1994) Role for supplementary motor area cells in planning several movements ahead. Nature 371:413–416

Tanji J, Taniguchi K, Saga T (1980) Supplementary motor area: Neuronal response to motor instructions. J Neurophysiol 43:60–68

Tanji J, Okano K, Sato KC (1987) Relation of neurons in the nonprimary motor cortex to bilateral hand movement. Nature 327:618–620

Tanji J, Okano K, Sato KC (1988) Neuronal activity in cortical motor areas related to ipsilateral, contralateral, and bilateral digit movements of the monkey. J Neurophysiol 60:325–343

Tanne J, Boussaoud D, Boyer-Zeller N, Rouiller EM (1995) Direct visual pathways for reaching movements in the macaque monkey. Neuroreport 7:267–272

Thaler D, Chen YC, Nixon PD, Stern CE, Passingham RE (1995) The functions of the medial premotor cortex I. Simple learned movements. Exp Brain Res 102:445–460

Thaler DE, Rolls ET, Passingham RE (1988) Neuronal activity of the supplementary motor area (SMA) during internally and externally triggered wrist movements. Neurosci Lett 93:264–269

Tokuno H, Tanji J (1993) Input organization of distal and proximal forelimb areas in the monkey primary motor cortex: A retrograde double labeling study. J Comp Neurol 333:199–209

Toyoshima K, Sakai H (1982) Exact cortical extent of the origin of the corticospinal tract (CST) and the quantitative contribution to the CST in different cytoarchitectonic areas. A study with horseradish peroxidase in the monkey. J Hirnforsch 23:257–269

Turner OA (1948) Growth and development of the cerebral cortical pattern in man. Arch Neurol Psychiat 59:1–12

Tyszka JM, Grafton ST, Chew W, Woods RP, Colletti PM (1994) Parceling of mesial frontal motor areas during ideation and movement using functional magnetic resonance imaging at 1.5 Tesla. Ann Neurol 35:746–749

Uematsu S, Lesser R, Fisher RS, Gordon B, Hara K, Krauss GL, Vining EP, Webber RW (1992) Motor and sensory cortex in humans: Topography studied with chronic subdural stimulation. Neurosurgery 31:59–71

Vaadia E, Benson DA, Hienz RD, Goldstein MH (1986) Unit study of monkey frontal cortex: Active localization of auditory and of visual stimuli. J Neurophysiol 56:934–952

Van Essen DC (1997) A tension-based theory of morphogenesis and compact wiring in the central nervous system. Nature 385:313–318

Vogt C, Vogt O (1919) Allgemeinere Ergebnisse unserer Hirnforschung. J Psychol Neurol 25:279–461

von Bonin G, Bailey P (1947) The Neocortex of Macaca Mulatta. University of Illinois Press, Urbana, Illinois

von Economo K, Koskinas G (1925) Die Cytoarchitektonik der Hirnrinde des erwachsenen Menschen. Springer, Wien

Vorobiev V, Govoni P, Rizzolatti G, Matelli M, Luppino G (1998) Parcellation of human mesial area 6: Cytoarchitectonic evidence for three separate areas. Eur J Neurosci 10:2199–2203

Walker AE (1940) A cytoarchitectural study of the prefrontal area of the macaque monkey. J Comp Neurol 73:59–86

White LE, Andrews TJ, Hulette C, Richards A, Groelle M, Paydarfar J, Purves D (1997) Structure of the human sensorimotor system. I: Morphology and cytoarchitecture of the central sulcus. Cereb Cortex 7:18–30

Woods RP, Grafton ST, Watson JD, Sicotte NL, Mazziotta JC (1998) Automated image registration: II. Intersubject validation of linear and nonlinear models. J Comput Assist Tomogr 22:153–165

Woolsey CN, Settlage PH, Meyer DR, Sencer W, Pinto HT, Travis AM (1952) Patterns of localization in precentral and "supplementary" motor areas and their relation to the concept of a premotor area. Res Publ Assoc Res Nerv Ment Dis 30:238–264

Wree A, Schleicher A, Zilles K (1982) Estimation of volume fractions in nervous tissue with an image analyzer. J Neurosci Meth 6:29–43

Yousry TA, Schmid UD, Alkadhi H, Schmidt D, Peraud A, Buettner A, Winkler P (1997) Localization of the motor hand area to a knob on the precentral gyrus. A new landmark. Brain 120:141–157

Zilles K, Schlaug G, Matelli M, Luppino G, Schleicher A, Qü M, Dabringhaus A, Seitz RJ, Roland PE (1995) Mapping of human and macaque sensorimotor areas by integrating architectonic, transmitter receptor, MRI and PET data. J Anat 187:515–537

Zilles K, Schlaug G, Geyer S, Luppino G, Matelli M, Qü M, Schleicher A, Schormann T (1996) Anatomy and transmitter receptors of the supplementary motor areas in the human and nonhuman primate brain. In: Lüders HO (ed) Supplementary Sensorimotor Area. Lippincott-Raven, Philadelphia, pp 29–43

Subject Index

www.ingramcontent.com/pod-product-compliance
Ingram Content Group UK Ltd.
Pitfield, Milton Keynes, MK11 3LW, UK
UKHW021334070726

13610UKWH00011B/64

* 9 7 8 3 6 4 2 1 8 9 1 1 1 *